Growing Cannabis
&
Cannabis Cookbook

A Complete Guide on How to Grow Marijuana Indoors, Make Delicious CBD and THC Sweet Edibles and Cannabis Edible Entrees to Heal Everything from Anxiety to Chronic Pain

By Tom Gordon

Table of Content

GROWING

CANNABIS

A Beginner's Guide to Cultivating and
Consuming Marijuana for Recreational or
Medical Use

By Tom Gordon

INTRODUCTION

Marijuana used to be considered the product of nightmares. Everyone's favorite cartoon characters from the 1980s spoke about the dangers of marijuana. But within the last twenty years, more research than ever has been done on cannabis and the results tell a different story. While it remains true that there are ill side effects when it comes to marijuana use, the benefits have been shown to far outweigh the negatives when considered across the board, especially when considered next to caffeine, nicotine, and alcohol, all of which have been perfectly legal for years. All of the benefits of cannabis consumption have pushed North America closer and closer towards coast-to-coast legalization. In the United States, there are eleven states that allow recreational marijuana use over the age of twenty-one and thirty-three states that allow marijuana use for medical purposes. Canada legalized marijuana in October, 2018, and added more than $8 billion dollars to their GDP as a result. With legalization spreading, many people are thinking about cultivating their own plants but don't know where to start.

That's where this book comes in. In *Growing Cannabis: A Beginner's Guide to Cultivating Marijuana*, you have everything you need to know to get started growing your own cannabis. We'll first start by taking a look at the history of cannabis, its side effects, and the benefits of using it. We'll follow this up with a look at the plants themselves to understand how marijuana comes in strains and is roughly divided into two types: sativa and indica.

Following this deep dive into cannabis, we'll look at where we grow our plants. Growing marijuana indoors leads to different results than growing it outside. Which of these techniques is right for you will be left for you to decide. Once we have our setup in place, we'll turn our attention towards the seeds and see how we germinate them. This will bring us to look at what are called "mother plants" and how we use these plants in cloning. Don't worry, it's not nearly as sci-fi as it sounds.

From cloning, we'll move into harvesting to learn the best practices for harvesting the fruits of our labor. The book will then close out with

a discussion on how to maintain your crops and prevent pests, fungus, or molds, as well as some tips on avoiding the common mistakes new growers seem prone to making. By the end, you will have all the knowledge you need to grow your own crop and avoid the costly pitfalls that trap less-prepared gardeners.

CHAPTER ONE

WHAT IS CANNABIS?

While marijuana certainly made its biggest impact in the last few decades, mankind has had a relationship with the cannabinoid plant family since well before recorded history. In fact, evidence dates man's discovery of the cannabis plant back to the Stone Age, around 8,000 BCE, with some scholars pointing even further back to 10,000 BCE. The history of marijuana is utterly fascinating—it takes you throughout history and all over the globe. In this chapter, we'll briefly look at the history of the cannabis plant, but our attention is primarily focused on the history of cannabinoid cultivation, which has changed drastically within our lifetimes and continues to do so as the laws become more lenient.

The reason that marijuana has exploded in popularity in the past decade is due entirely to the increased research and attention on the plant. The main psychoactive component in cannabis is called tetrahydrocannabinol (THC), which has proven to be a much more interesting chemical than previously thought as much of the original research that showed us marijuana was harmful was in fact quite flawed. One of the most famous examples of this is the idea that marijuana kills brain cells. The study that came to this conclusion did so by strapping a gas mask onto a chimp and blasting it with marijuana smoke. The problem with this experiment's setup is that researchers didn't give the subject air, just smoke. Brain cells start to die when they cannot get air, regardless of what kind of smoke is being inhaled. Results like these have been proven to be scare tactics founded on faulty science, and there are actually many benefits to using marijuana responsibly. However, honest

science has shown that marijuana does have some negative side effects. In order to give you a fair and accurate look at this controversial plant, we'll look at both the benefits and the risks of marijuana in this chapter.

A History of Cannabis Cultivation

While Asia has been identified as the location of origin for marijuana, it is uncertain whether the plant came out of China, Central Asia, or South Asia. In addition, many strains find their genetic roots trace back to Afghanistan and the Hindu Kush mountains (which is the reason why many strains have Kush in their names). Nevertheless, the earliest sign of cannabis cultivation comes from China between 8,000 BCE and 6,000 BCE. Evidence points to the Chinese of this period having a strong relationship with the plant, using it in many aspects of their daily lives. The hemp from the plant was used in textile production, the seeds were used as food, and even the oil seems to have been harvested for use. The Chinese are considered the inventors of farming, and one of the earliest emperors, Shennong, is credited as the inventor of many early farming tools. Shennong published one of the oldest known medical texts, and in this text, lists cannabis as a medicinal herb.

What this tells us is that from the very beginning of mankind's history with cannabis, we have been aware of its many benefits from its use in industry to its medicinal value. In an intriguing piece of contemplation, Carl Sagan considered the possibility that cannabis was the earliest crop cultivated by man. If this were so, then marijuana directly contributed to the invention of farming and our subsequent agricultural revolution. Whether Sagan's suggestion is true is yet to be seen, but the idea does present just how important cannabis cultivation has been to mankind.

Scythians, a nomadic culture from Central Asia, traded for cannabis seeds around 1,500 BCE and thereafter began to cultivate their own crops. In 600 BCE, Persia discovered cannabis. They referred to it as "bhang," or "the good narcotic." Persia's discovery of cannabis brought the plant into contact with the rest of the known world, as hemp rope can be traced to Russia in this same period. Persia's conflicts with Greek introduced the plant into post-Homeric Greece. While it is not clear exactly when or where the Roman Empire encountered cannabis, the plant made its way into their society in roughly the same period that the

calendar changed over to the common era (CE). The Vikings failed to cultivate the herb in Norway and Iceland upon first discovery, but in time they managed to produce strains that could survive the cold. As a possible consequence, the Vikings may bear responsibility for bringing cannabis to North America. At the same time, the Arabs were discovering the plant. By the 14th century, cannabis was in Africa. If the Vikings hadn't brought marijuana to North America prior, then the Spanish shipped it to North America and Chile in the late 16th century. With cannabis having spread across the globe, this brings us into issues of modern cultivation.

In what is a complete mirror of the 1980s, one of the first rules related to cultivating marijuana was set by King Henry VIII, who fined farmers that would not grow the plant. Henry VII recognized the high value of hemp fibers from the plant and sought to grow as much of it as possible to fuel trade. In America, an early law required farmers to grow Indian and English hemp in their fields. Hemp cultivation in the United States was particularly fruitful thanks to ideal growing conditions. Later, the 19th century saw cannabis cultivated more for medical reasons than for its hemp. However, despite America's early relationship with the plant, laws first aimed at restricting the use of opium would grow to encompass alcohol and marijuana use with the cultivation of the latter becoming illegal. The controlled substances act of the 1970s further criminalized the substance and Nixon's War on Drugs effectively destroyed the public image of marijuana.

Despite the legalities involved, people continued to grow cannabis all over the world and even in America. However, the main source of cultivation came from strict laws designed to produce hemp or provide researchers with enough of the plant to carry out research. That research, which rediscovered marijuana's possible benefits, has led to a resurgence of cannabinoid popularity, and this popularity, backed by science, has led to changes in many of the laws around marijuana use and growth. Depending on where you live in North America today, it may still be illegal to possess or grow cannabis. Or, it may be legal to possess but not to grow cannabis. It may also be legal to grow these plants up to a certain amount, or you may even be able to get permits to grow crops for distribution to pharmacies. One thing is becoming clearer every day: Marijuana cultivation is on the rise and the laws are changing to reflect its growing popularity. Still, when it comes to growing your own

marijuana, always check the laws of your country, state, and county to ensure that you stay on the side of legality.

Side Effects of Using Marijuana

Many advocates for marijuana use claim (and even believe) that there are no harmful or negative effects associated with cannabis. It would be amazing if such a claim was true, but the reality and science indicates that there are negative effects associated with marijuana use and users should be aware of them. However, it is worth noting that some of these negative effects may, depending on the user's circumstances, in fact, be positive. For example, some people may see increased appetite as a negative side effect, but for someone having

trouble with nausea or eating (due to, say, the effects of chemotherapy), this side effect would be a positive.

When smoking or consuming marijuana, there are a number of associated short-term effects collectively referred to as the "high." These effects include distorted perception, and the user's sights, sounds, time, or touch may seem slightly off. Issues with coordination, memory, and learning all present as side effects as well. Likewise, the user may find reductions in their critical thinking and problem solving abilities. An increased heart rate is also common. These effects in and of themselves are quite often the user's goals. For example, the distorted perception can make viewing movies or listening to music a deeply engrossing or even spiritual experience. However, as enjoyable as these situations may be, they are still effects that the user should be aware of.

In some cases, anxiety, fear, paranoia, or even panic may be found among the short-term effects experienced by the user. The mental healths of the user and the emotions that they are feeling when they partake in marijuana have a strong influence on how the user perceives the experience. This can lead to a "bad trip" or even "greening out," in which the user experiences an upset stomach or even vomit. More often than not, this reaction is due to the psychological experience the user is having and not the physical experience of their body.

The active chemical in cannabis, THC, attaches to cannabinoid receptors found on nerve cells in the brain. The THC then has an influence on how these cells function. Different parts of the brain have different amounts of these cannabinoid receptors. Some have many, while some have none. Unsurprisingly, the areas of the brain in which cannabinoid receptors are concentrated are those that deal with the experience of pleasure, memory, thought, concentration, perception, or movement. High doses of marijuana can cause these areas to function poorly, and in the worst experiences, this can mean hallucinations, delusions, or disorientation.

Meanwhile, while THC is attaching itself to the user's cannabinoid receptors, the effect of introducing a new chemical into the body leads to an increased heart rate. When the heart starts to beat quicker, there is a corresponding drop in blood pressure. Use of marijuana can see your heart rate increase between 20%-100%. This change in the functioning of the heart leaves users at a heightened risk of heart attack during the

first hour of smoking. This increased risk is not high enough to be a major concern to the average user, but anybody that is combining medications or dealing with heart-related medical issues should be wary about consuming marijuana and consult with a physician beforehand.

Another issue which was brought into light in 2017 is marijuana's relationship with the body's bones. Studies have found that users who smoke large quantities of marijuana on a regular basis have a reduced bone density compared to light or nonsmokers. As a result, heavy marijuana smokers have greater susceptibility to fractures or broken bones. The study which discovered this also found that those users have a lower body mass index, which may be the reason for the lower bone density. However, a similar study from 2017 did not find any connection between bone density and marijuana use, which means that this finding is still up in the air and requires more research.

Smoking marijuana causes the lungs and throat to burn and can cause fits of heavy coughing. Users have for years claimed that there is no harmful effect to smoking marijuana, but a study in 2019 found that regular intake of marijuana smoke can cause similar respiratory issues to that of tobacco. A persistent cough, frequent chest illnesses, and an increased risk of lung infections are all possible side effects. Many marijuana smokers intake less smoke (volumetrically) than the average tobacco user and this can be one reason we see fewer lung issues in marijuana smokers. It is not the chemical, THC, that causes this issue, but the act of taking smoke into the lungs. One clear issue in relation to marijuana smokers is the method through which many users inhale deeply and hold the smoke in their lungs. This practice is seen as a way to increase the amount of THC entering the body, but doing so also exposes the lungs to the carcinogenic properties of the smoke for a greater period of time. However, despite the similarities between smoking marijuana and tobacco, there has not yet been any confirmed research similarly linking marijuana use to increased risk of cancer.

One other area that we need to address is the use of marijuana during pregnancy. It is never a good idea to take substances when pregnant and marijuana is no exception. Studies have found that prenatal exposure to marijuana can result in children born with strange responses to visual stimuli, increased tremulousness, issues with attention and memory, and poor problem solving skills. If you wouldn't drink while pregnant, then you shouldn't smoke while pregnant either.

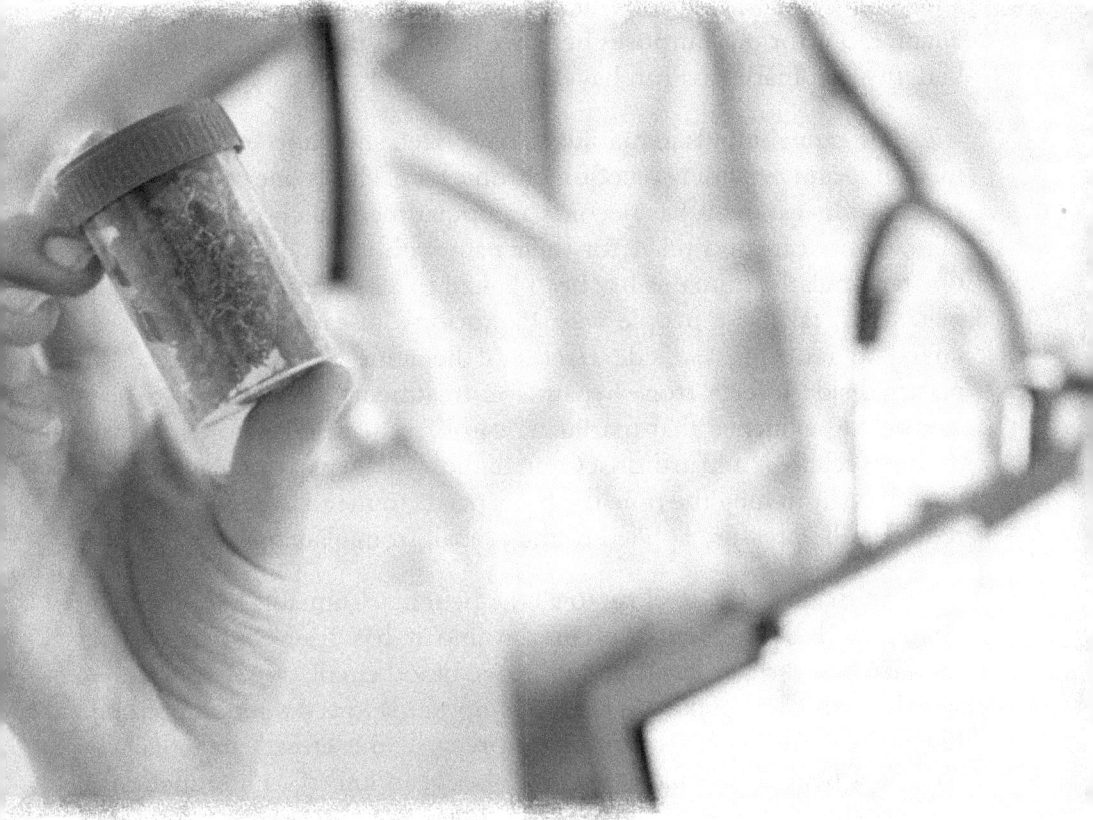

The Benefits of Marijuana Use

If everything was negative, there would be no push for the widespread legalization of marijuana. But as you will see, the benefits are many. One of the things that we should take note of here is CBD. CBD stands for cannabidiol and it is the second most common compound to marijuana after THC. While THC is the chemical which provides the "high," which can bring both positives and negatives, CBD provides no high. It does, however, offer many benefits with no evidence of harmful side effects (as of this writing). The presence of these two chemicals is important to note because growers may aim to produce marijuana with a low THC and high CBD, or a high THC and low CBD, or any

combination thereof. What this means for us is that the benefits associated with marijuana may be different depending on the particular strain discussed. We'll look at the importance of strains in the next chapter, but for our purposes here, we'll look at the overall agreed-upon benefits that marijuana can have.

The cannabinoids in marijuana have been shown to be of benefit in managing and providing relief for chronic pain. This is one of the primary reasons why cannabis has become a popular medical choice. With CBD oil, patients can find relief for their pain without having to turn to an opiate. Similarly, cannabis has been linked to a reduction in tremors and pain when taken by people with Parkinson's disease. It has also been linked to reducing the side effects of hepatitis C and the impact of negative side effects from hepatitis C treatment. There has also been compelling evidence that marijuana can offer pain relief to those with multiple sclerosis and arthritis. Cannabidiol (CBD) in particular has been linked to shortening the time the body needs to mend broken bones and reducing the pressure applied to the eyeball from glaucoma.

But that's not even close to all the benefits. Cannabis has an impact on the body's production of insulin and it has been shown to help stabilize blood sugar levels and improve blood circulation. Likewise, we saw that cannabis intake can lead to temporarily lowered blood pressure. For those suffering from high blood pressure, this effect can actually be seen as a benefit. Cannabis use has also been linked to a reduction in weight. This finding is linked to the findings on diabetes since the change in the body's insulin helps the body to more efficiently use the calories that are consumed.

There is a ton of research showing that cannabis can help to fight certain kinds of cancer. This is the most controversial of the benefits and should not be blown out of proportion. Marijuana use has been linked to an improvement and reduction in cancer in patients but it is not in itself a cure for cancer. Taking smoke into the lungs can still cause lung cancer, even if the smoke comes from marijuana and not some other substance. However, along with these intriguing findings, there are many first-hand reports about the helpfulness of marijuana consumption in limiting nausea from chemotherapy.

Cannabis has been used to help battle alcohol addiction, though much work is still needed in this area. It has also been used to help

people with Crohn's disease or ulcerative colitis to reduce inflammation in the intestines. There has been a lot of research showing that cannabis can help in the treatment of PTSD when used in conjunction with a mental health specialist. Likewise, cannabis has been shown to reduce anxiety in many users. However, people that use marijuana to treat anxiety have also reported the effects turning against them and making them more anxious, so this particular area can be either beneficial or harmful. This finding is also repeated in those looking at depression. Cannabis has also been shown to help people with ADHD focus better and has been considered as an alternative to traditional drugs such as Ritalin or Adderall.

Whether it is your mind or your body, there are many ways that the use of marijuana can be beneficial to the user. The most intriguing part of these benefits is that marijuana offers a far safer alternative to some of the scarier chemicals used medically today such as Adderall and opiates. Still, despite the many potential benefits, it is always best to consult with a medical professional before using marijuana, and it is always a good idea to start slow and see how your body reacts to cannabis.

Chapter Summary

- Humanity's relationship with cannabis goes back as far as 10,000 BCE.

- The earliest signs of cannabis cultivation date to between 8,000 BCE and 6,000 BCE and come from China.

- Evidence suggests that the Chinese used cannabis on a daily basis as both a food and a medicine.

- The Chinese are also the inventors of farming, which has led some to suggest that cannabis was humanity's earliest cultivated crop.

- Cannabis spread out from Asia, making its way all over the world from Europe to Iceland, North America, and Africa.

- One of the first laws relating to cannabis was a demand by King Henry VIII to grow more. One of America's earliest laws relating to cannabis also required farmers to cultivate the plant.

- Cannabis only gained its illicit status in the 1900s due to anti-substance campaigns in the 1920s. Recent years has found its cultivation and use resurge as new research has led to changing attitudes about marijuana.

- Short-term side effects of smoking marijuana include distorted perception, issues with memory and learning, issues thinking or solving problems, and increased heart rate.

- Some people may experience fear, paranoia, panic, or anxiety after smoking cannabis.

- THC, the active chemical in cannabis, attaches to receptors in the brain to cause the effects associated with the "high" of marijuana.

- As THC attaches to the brain, the user's heart rate increases between 20%-100% leading to reduced blood pressure. If you have any heart issues, consult with a doctor before using marijuana.

- Smoking cannabis also exposes the user to the possibility of developing issues with the lungs. However, there has yet to be any confirmed link between use of cannabis and lung cancer.

- Cannabis should also not be used by pregnant women, as there are links between cannabis use and learning disabilities.

- Another active ingredient in cannabis is CBD, which helps to reduce pain and provide many benefits to users.

- High CBD strains are recommended for use in pain management as an alternative to opioids.

- Cannabis has also been linked to helping reduce tremors and pain in people suffering from Parkinson's, reduce the side effects of hepatitis C treatment, and shorten the time it takes bones to heal.

- Cannabis has also been shown to have a positive impact on how the body uses insulin, as well as help in weight reduction for these reasons.

- There is strong evidence that cannabis use can help in preventing or battling cancer, though these findings are often treated like a magic cure. While cannabis helps in managing cancer, it does not and should not replace proper medical treatment.

- Marijuana has been used to help battle alcohol addiction, reduce anxiety and depression, and help people with ADHD to focus better.

- Every day, there are more findings about the beneficial effects of cannabis. However, there are also more findings about possible downsides to cannabis use. If you have any medication conditions, do your research and speak with your doctor about cannabis use prior to beginning.

In the next chapter, you will learn why genetics are important in growing and cultivating cannabis. You will also learn what we mean when we speak about different strains of marijuana. Finally, we'll look in depth at sativa, indica, hybrids, and ruderalis, the main varieties that the cannabis plant comes in. Each has its own use and attributes that you will want to consider before seeding your first crop.

CHAPTER TWO

UNDERSTANDING THE PLANT

All of the benefits and side effects that we looked at in the previous chapter are dependent upon the particular strain of marijuana a person uses. Some strains will have stronger pain management effects, for example, while others will be more "heady" and relate to the effects that impact, or impair, the user's sense of perception. In this chapter, we'll see what exactly a strain is and how the genetics of our plants plays a role in determining the strain. There are two key categories within which most strains fall: sativa and indica. We'll explore what these descriptors mean and the rise of hybrid plants that combine traits from both types.

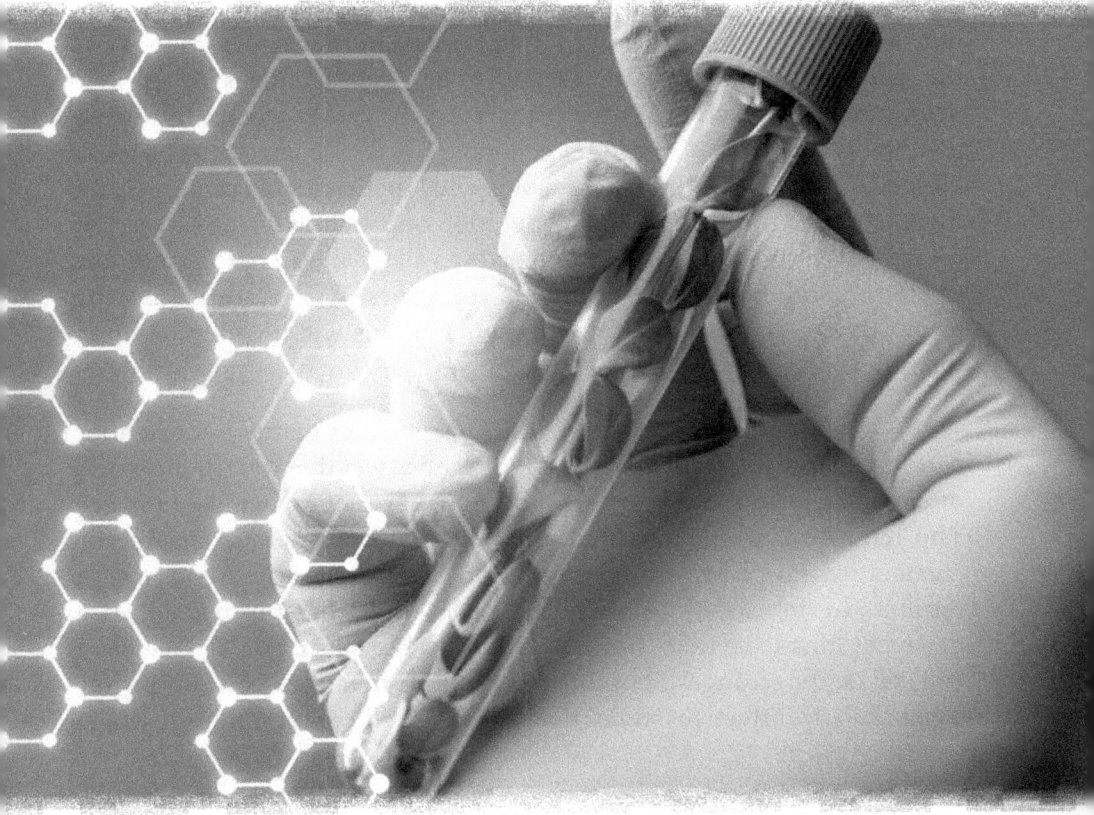

Why Do Cannabis Genetics Matter?

We know how important genetics are when it comes to human beings. Our genes determine whether or not we are susceptible to certain illnesses, what we'll look like, and whether we'll be male or female when we're born. That last trait is determined by whether or not we have two X chromosomes (female) or an X and Y chromosome (male). The X chromosome in a human has around 800 different genes, while the Y chromosome contains only 70. The X chromosome contains about 150 MB of information while the Y chromosome has 3,300 MB. All of this data results in the amazing variety of human beings that the world sees today.

Cannabis is remarkably similar to humans in many regards when it comes to genetics. The plant has an X and a Y chromosome as well, though each chromosome contains between 800-850 MB of genetic information. Every single part of the plant, from how its leaves looks to the percentage of THC it has and even how it smells, is entirely determined by which genes the plant has. Growers work hard to breed plants together to isolate certain genetic traits they consider to be ideal. However, there are so many possible genes that cannabis can have that there is now a ton of research being done on creating a genetic map of cannabis. The goal of this genetic map is to allow researchers around the world to combine their data on traits and genes. With research moving in this direction, more and more designer strains will be popping up in the future.

When it comes to growing marijuana, genetics matter because they determine how the plant will grow, look, and affect the user. In the world of cannabis, plants come in male and female varieties just like humans. The female plant contains the bud, which is what people smoke. To create a hybrid plant, you take the male plant of one kind and have it fertilize the female of another kind. The seeds that the female plant produces now have traits from both types. The traits of the female plant will still be the strongest, but through multiple iterations you could grow a particularly effective strain that combines the best of both origin strains.

With that in mind, let's look a little more at what we mean when we talk about the various strains of cannabis.

What Are Strains?

The cannabis plant began in Asia but then spread all throughout the world. One of the pieces of history that we saw in the last chapter was how the Vikings had acquired the plant and were able to get it to grow in their colder climate. In doing this, the Vikings essentially created their own strain of marijuana. Cultivators across the world have had to breed their plants to grow in a variety of different climates. As these plants grow, develop, and continue to mate and reproduce, they turn into what is known as strains.

Strains are essentially variations in the genetics from one cannabis plant to the next. Before, strains arose naturally because of the needs of the individual growers to acclimatize the plants. Today, however, with the advent of indoor gardening and the ability to control any plant's individual environment, strains have become much more about what genes a particular plant was grown to have. The global cannabis industry is expected to be worth more than $31 billion by 2021. This means that there is a lot of money involved in growing and where the money goes, skill follows. Skilled growers have now created thousands of strains of marijuana.

There are three general categories that strains fall into: indica, sativa, and hybrid. Each of these strains tells the user something about the effect that the strain will have when smoked (and we'll explore these each more in a moment). But within each of these categories, the various strains may have major differences, particularly when it comes to the compounds found in each strain. We looked at THC and CBD briefly but they are far from the only cannabinoids found in marijuana.

CBN helps in reducing the effects of neurological conditions such as epilepsy or seizures. THCA is similar to THC but is most useful in reducing inflammation and reducing neurological conditions associated with Parkinson's or ALS. CBG has been tied to reducing anxiety and other psychological issues such as OCD, PTSD, or depression. The proportion of these compounds present in the plant differs between strains.

Another area in which there is a wide variety is in the terpenes that are present and the quantity of them. Terpenes are a compound that naturally occurs in the plant and is primarily related to the way the plant smells. Marijuana is often described as flowery, skunky, or fruity in smell. Clearly, these descriptions would appear to be contradictory, since who ever heard of a fruit that smelled like a skunk. But it is the terpenes present in the strain that determines how the plant smells and creates this variety. It should be noted that while terpenes are most strongly associated with the smell, they also have effects on how the strain functions. A quick look at some of the most common terpenes will help enlighten this point.

Bisabolol is thought to help reduce inflammation and irritation of the body and has some microbial and pain-reducing effects as well. It

also smells like chamomile. Limonene, on the other hand, smells like citrus and is thought to improve mood and reduce stress. Eucalyptol smells like eucalyptus, reduces inflammation, fights bacteria, and leaves the user feeling refreshed. Pinene smells like pine and helps ease nausea and pain. Linalool smells of flowers and helps boost mood. These are just a few of the many terpenes that are found in cannabis, each of which can be found in certain strains and may even help to give strains their names, such as Sour Diesel, which gets its name from its pungent aroma.

Let's take a look at a handful of strains. We cannot even begin to cover them all—that would easily take a book a dozen times the size of this one—but we'll look at some of the most common ones.

Sour diesel is known for its energizing and mood-lifting effects. Depending on the kind of work at hand, it can help provide a burst of productivity to the user's day and has been tied to the reduction of pain. Bubba Kush, on the other hand, helps relax the user and makes them sleepy. Hence, it is recommended to help fight insomnia. Bubba Kush is also linked to a reduction in pain and stress. A similar strain is Granddaddy Purple. Granddaddy Purple helps to induce sleep as well, but it also increases the user's appetite, which can be extremely useful for those whose medical issues result in a lack of appetite or nausea.

These strains are recommended for certain results because they have been grown to have the effects they do. Those mentioned above have been around for a while now, and this means that you can expect a certain level of consistency when using them. In contrast, newer strains or hybrids will show more variation between crops. Finding the strain that is right for you may take some time but sites like weedmaps.com and leafly.com compile information on the strains out there alongside what traits they have and even ratings from staff and users.

Figuring out what strain is right for you first starts by exploring the categories that were mentioned above: sativa, indica, and hybrid. Keep in mind, however, that individual strains of any one category may have effects that are not associated with that category. For example, indicas are seen as relaxing but a particular strain may in fact leave the user feeling more energized. The effects of marijuana should be viewed on a strain by strain basis, with the categories used to help the individual to narrow down their search for the right strain.

KANDY KUSH
50/50

Sativa

Sativas are known for the "heady" high that they provide when used. They tend to leave the user feeling invigorated. They're like the energy drink version of cannabis, only without all the sugar and dangerously high levels of vitamins. Many people find that sativas help them feel less anxious. They have also been linked to an increase in creativity in individuals, with people reporting that they feel they do

some of their best creative work after using a sativa. Another link that has been seen is an increased ability to focus after using a sativa.

The sativa family comes from climates that are hot and dry. It prefers longer days, which means lights needs to be lit longer if growing indoors. Its origins are traced back to Africa, Central America, and Southeast and Western Asia, where the climate perfectly fits its needs. The plant itself tends to be tall but thin, with leaves that almost look like the fingers of a spread hand. They can pass twelve feet in height with the right conditions, but their height also means that they grow slower than other varieties. Because sativas are more of the "heady" experience, it is common for them to have high amounts of THC but low amounts of CBD.

Since sativas are stimulating, they are best used during the day as users may find it difficult to sleep after consumption. This makes the sativa most useful for individuals who work in the creative fields or who want to experience the psychological effects of cannabis over the physical benefits.

Indica

Indicas are most strongly associated with what is called a "body high." These strains are known for their relaxing properties and thus are among the most popular for battling insomnia, inflammation, nausea, and pain. A common use for indica is in reducing the side effects of chemotherapy. People undergoing chemotherapy often are so nauseated that they spend hours after each session vomiting. Use of an indica before a session has shown great results in reducing these side effects and helping patients keep their appetites and improve their mood.

Indicas find their origin from the plants of Afghanistan, India, Pakistan, and Turkey. In doing so, they are particularly well suited for growing in climates that are harsh or dry. They grow particularly well in the turbulent climate of the Hindu Kush mountains, after which many strains have been named (Purple Kush, Granddaddy Kush, Bubblegum Kush, the list goes on and on). Indica plants are much shorter than sativa plants. They tend to be short, stocky, and densely covered in bushy greenery. Whereas the leaves of the sativa are finger-like, the leaves on

an indica are quite broad and more in line with what people think of when they picture a pot leaf. They grow much faster than sativa does, and they also tend to produce more buds, both of which make them a favorite amongst those who grow for financial reasons.

Indica is best consumed at night or when you are planning to stay at home and just relax. There is a common depiction of the cannabis user in the popular media as being too lazy to get off the couch. This image contrasts the effects of sativa but perfectly fits those of indica. The indica plant has much more CBD than the sativa and likewise also a lower ratio of THC, making it the more effective of the categories for medical purposes. However, this has begun to change due to the rise in hybrid strains in the last decade.

Hybrids

Hybrids are an interesting category of cannabis. They almost exclusively need to be created by growers, as the chances of a hybrid forming naturally in the wild is almost entirely impossible due to sativas and indicas preferring different climates. However, within a controlled growing operation, such climate variables can be accounted for to allow these disparate types of cannabis plants to interact with each other. Hybrids are formed by breeding an indica with sativa. Which category will be the stronger of the two depends on which is used to supply the female fruiting plant. The seeds produced are dominated by the genes of the female but with the genes from the other (male) brought into place. Because of this, hybrids have an especially large degree of variety as a category and even within an individual hybrid strain.

The appearance of a hybrid is entirely dependant upon the parent plants which were used to create it. The amount of THC or CBD they have is also dependant on this factor. However, hybrids are primarily created in order to increase the THC percentage of an indica or to increase the CBD percentage of a sativa, though the exact percentage of THC or CBD that a hybrid will have is a factor that cannot be accounted for in a generalization. This carries through to the effects most commonly associated with their use, as well as when they are best used.

Ruderalis

Any discussion on cannabis plants would be remiss not to make a quick mention of the ruderalis. This kind of cannabis plant grows quickly and adapts well to extreme environments. They're small, rarely more than a foot tall, which makes them able to go from seed to harvest in roughly a month's time. They have almost no THC and only a little more CBD, though most likely neither is enough to produce effects in the user and so the plants are not used for recreational or medical purposes. However, the ruderalis is able to breed with sativa and indica plants. Their ability to breed across the board, along with their short growing time, make the ruderalis well suited for those individuals who want to experiment with creating hybrids, shorten the length of a strains growth cycle, or try to isolate particular genes for further use.

Chapter Summary

- The genetics of cannabis is shockingly similar to that of humans, even down to having male and female plants.

- The genes that a plant has are determined by the genes of its mother plant.

- The female cannabis plant is fertilized by the male plant and then produces seeds. The female plant is also the one to grow the bud which people smoke.

- Everything from the THC to CBD ratio to the smell of the plant is determined by the genes it has.

- Strains of cannabis are plants which share the same genes. Every strain has different genes and this allows for a wide variety of effects across thousands of strains.

- Strains can be divided into three general categories: sativa, indica, and hybrids.

- Other active ingredients in cannabis include the cannabinoids CBN, THCA, and CBG, all of which are found in different ratios depending on the genetics of the strain.

- CBN helps in reducing neurological effects such as epilepsy.

- THC reduces inflammation.

- CBG reduces anxiety and other psychological distressors.

- Many strains are named for the smell they have. Cannabis gets its smell from the terpenes which are present, another factor which is determined by genetics.

- Besides providing marijuana with its smell, terpenes may also have beneficial effects such as lifting mood or reducing inflammation.

- Experienced growers work hard to create powerful strains which provide the most beneficial effects possible.

- Older strains are more predictable, because they have had a long time of developing the same genetics from generation to generation. In contrast, newer strains may see a lot of variation between crops.

- While strains are split up into sativa and indica, a particular strain may actually have effects more commonly associated with the other kind, such as an indica that has the energizing effects of a sativa.

- Sativas are associated more with the "high" of marijuana, producing the most notable psychological effects.

- Sativas grow tall and slow, enjoying hot and dry climates with long days.

- Sativas are high in THC but tend to have low CBD. They are best used during the day because of their energizing effect.

- Indicas provide the most body-heavy effects, such as reducing pain or inflammation.

- Indicas flourish in harsh climates. They are shorter, faster growing, and produce more buds than sativa plants.

- Indicas are known for their high CBD ratio and a subsequent lower amount of THC.

- Hybrids are combined by breeding a sativa with an indica. Their preferred climate, appearance and effect will be determined by the parent plants.

- Hybrids are often created to raise the THC level of indicas or the CBD level of a sativa.

- A hybrid will have the strongest genes from the mother plant, leading people to categorize hybrids as either sativa-indica or indica-sativa with the mother plant's type coming first.

- Hybrids have the most variety in the genes of their strains.

- There is also a class of cannabis plant called the ruderalis. This plant does not produce enough THC or CBD to have an effect when consumed, but it can mate with either sativas or indicas, which makes it useful for growers looking to isolate genes.

- The ruderalis also grows the quickest of any type, from seed to harvest in roughly a month, which makes it ideal for combining with other plants to create quicker growth cycles.

In the next chapter, you will learn about the ways in which the growth of cannabis is affected by the environment it grows in. Cannabis is grown both indoors and outdoors, and you will see the pros and cons of both options in order to pick the optimal setup best for your specific needs.

CHAPTER THREE

WHERE TO GROW YOUR MARIJUANA

When it comes to gardening, there are two options available to us. We can either grow our crops outdoors and rely on the climate of where we live and the unpredictability of Mother Nature or we can design our crops to grow indoors, where we can control the climate and remove variability from the equation (if we are mindful and remove chances for infection or pests).

Cannabis is a particularly flexible plant, able to flourish both indoors and outdoors. Outdoor growers may choose to include a plant or two in their backyards among bushes or flower beds. They can also drop seeds throughout a forest and come back to find plants that survived to harvest. Of course, the primary method for outdoor growing is to plant a proper crop and oversee it through the growth cycle, which is what we'll be considering when we look at outdoor growing, but it is truly impressive just how versatile cannabis can be when it comes to growing.

Indoor growing is the more popular approach these days. Growing indoors allows the gardener to control the conditions the plant experiences. This allows for the conditions to be set just right for the most plentiful harvest. It also allows the gardener to reduce contamination from the environment or accidental fertilization from another strain. However, growing indoors presents its own challenges which must be considered as well. Let's take a look at the key differences between these two approaches and then consider them each themselves.

Growing Indoors versus Outdoors

Making the decision on where to grow your cannabis requires that you consider three primary factors. The first factor is the amount of control you have over the growing process. While you may naturally assume that more control is always a good thing, you will see this isn't quite true. The next factor, and the only factor that matters to many people, is the upper limit you are willing to spend on growing. This cost factor goes hand in hand with control; the more control you want, the more you can expect it to cost you. The final factor is the quality of the cannabis. We aim for the highest quality, but getting there will be determined by the other two factors, your skill level, and your luck.

Growing your cannabis indoors allows you to take over full control of the growing process. Cannabis doesn't just grow inside the house by sitting in a pot in the corner or next to the window. The cannabis plant requires that you consider its needs and build an area dedicated to its growth. If properly laid out, then this area will allow you to control everything about the plant's environment from the temperature to the lighting to the CO_2 production to the humidity. It will also allow you to easily get around the plant for harvesting, trimming, and upkeep.

Growing indoors often provides the best product. However, the power of a light bulb is nothing compared to the power of the sun. Indoor plants will produce fewer buds, and the plants will not be as strong. There are other issues involved in creating an indoor growing environment that passes as natural. One issue that will come up when we look at pests is the way that growing outdoors actually leads to fewer problems with mites when compared to growing indoors. This can make outdoor farming sound like a good idea, but growing outside only works when you live in a climate with the right amount of heat, sun, and humidity.

You need to consider the cost of operations when making your choice. However, you are going to be spending a decent amount of money regardless of the option you pick. It may cost more to start growing indoors because you will need to purchase climate control systems, but once that is in place, the cost levels out a little more. Because an indoor grow operation has a high turnover, the grower will always have something they can do to help the plants out. This eats up

a lot of time. But growing outside takes a lot of time to plant a crop, and then crop maintenance will require several hands. In addition, depending on the size of the crop, you may need to hire more hands when it is harvest time. This is because outdoor crops work by a seasonal basis and so everything is ready to harvest at the same time. The time and labor spent can quickly run up the cost of growing. If you are concerned about how much it will cost you to grow, it is a good idea to sit down ahead of making your decision and write out a budget and plan.

It is also a good idea to consider the way that each style of growing makes income. Indoor farming has a high startup cost but allows you to control the breeding of plants to make unique and high-end strains that are more potent. Indoor growing also harvests many times throughout the year, and so your income trickles in periodically. However, indoor growing also takes a lot of electrical energy since your lighting will replace the power of the sun. This cost in energy is subject to change, with almost every projection of the future saying that the price of energy will go up. An outdoor crop takes barely any energy at all and it produces much more buds. You might only be making an income on a seasonal basis, but the increase in yield means you may make more money off of it. However, the THC amount will typically be lower, and so it will not sell for as much as the indoor cannabis does.

Indoor growing leads to the best product. All of the highest quality strains are grown indoors, without fail. Having the ability to control the growing climate and speed up the breeding process has led to some amazing strains. Along with controlling the climate, you have the ability to control the CO_2 level in your growing area. Higher CO_2 levels lead to faster growth and higher THC levels. By remaining indoors, the plants aren't exposed to any harmful environmental factors like wind or rain that could harm them. The buds on your plants only risk one thing harming them: You. If you handle your plants properly, and prevent infection or infestation, then the plants shouldn't have any troubling degradation.

Where the indoor plants are completely under your control, the outdoor plants are under the control of Mother Nature. They typically don't look nearly as pretty as indoor plants do, with a product that isn't as appealing aesthetically, despite their chemical components being perfectly fine. There are many people who prefer the look of a natural,

outdoor grown bud and so this may work to your advantage if you are crafty. There is an idea that outdoor grown cannabis is always ugly. This comes from the fact that many outdoor growers are doing so outside of the law, and so the plants and the yield are cared for quickly rather than carefully.

However, a change has slowly been impacting the market. With the legality of cannabis, there are now many farmers using greenhouse techniques to combine aspects of both indoor and outdoor growing. This is by far the most expensive option and it is yet to be seen if these growers will be able to stay afloat on the income it provides. Time will tell whether or not this technique should be explored in more depth. For now, let's turn our attention towards outdoor growing for a few moments.

Growing Outdoors: The Best Places to Grow

The best part of growing outdoors is the many options available to you in terms of location. You could grow just about anywhere you can imagine. However, there are some locations that prove to be the most popular. There are also some concerns that should be addressed when picking your location, such as security.

Many, many different places have been used as grow sites over the course of human history (and especially as of recently), and this means that there are a lot of voices sharing their experiences. The best place that you can grow your cannabis is in your own garden. This gives you access to it whenever you want. However, if you have an open garden then it can be very easy for people to come and steal your product. This is unfortunately a more common problem for growers than you might imagine. More security is offered by growing on a balcony or on a roof terrace. However, the visibility of these plants may not be appreciated by your neighbors, and strong or persistent wind can be damaging to the plants. Fields are great locations for growing sites because they are open to lots of sun, but they certainly make your harvest noticable and this means a need for more security. Many growers who focus on smaller crops consider forests to be ideal locations. If it is off the beaten path and hence offers a much higher level of security, but it can be more difficult for plants to receive the proper amount of sunlight and water. Regardless of where the grow site is, being close to a river means the soil stays wet, and this can help your plants to stay properly hydrated.

Which of these locations is ideal depends on whether or not they are able to give your plants access to the environmental elements they need the most. Cannabis requires a lot of sunlight, with the various strains liking more or less on an individual basis. It also helps to have water nearby and to ensure that there is a breeze, but not one so powerful it can damage the plants. Soil that is rich in nutrients or that you can make rich yourself is almost (but not quite) a must. You also need to have easy access to continue caring for them throughout their

growth. But one of the biggest issues is security. Thieves are common in this industry, and you do not want to lose out on some major money because you hadn't considered the threat they pose.

Even though growing outside means relying on the whims of nature, you are still able to take over control of some of the elements. For example, if you aren't able to find soil that is naturally rich, you can purchase your own. Or, if you aren't able to find a location close to water, you can set up a watering system to compensate. The primary elements that make a location ideal is how much sunlight it gets, how much wind it is exposed to, and whether or not you can keep the plants protected from outsiders.

Growing Outdoors: Growing Calendar

Because growing outdoors ties you to the changing seasons, it means there is a very specific schedule from which you will be operating. Cannabis requires a certain amount of illumination each day, or photoperiod, so it can move from the vegetative stage into the flowering one. The following calendar is based on what is best for growers located in the Northern Hemisphere.

The best time to germinate your seeds is at the start of April, but it can be done in April, May and June. It is recommended that you germinate your seeds indoors at this period prior to heading outdoors. Depending on when you first started the germination process, you will be ready to move your seeds to their outdoor location between May to July. You can expect to need to top the plants between June and the start of August. Pruning and clean up will take from August to September. Then, if everything previous has gone well, and there have been no major disasters, the harvest season begins halfway through September and can last as late as mid-November.

It should be noted that where you are located affects these dates, even within the Northern Hemisphere. For example, Californian farmers will have a longer period to work since the weather is always so warm there. Those in Washington, on the other hand, will find that they have to wait until later in the season to germinate if they want their plants

to receive enough sunlight to stay healthy and they'll even need to harvest earlier because the weather turns so drastically.

You can use the Spring and Fall Equinoxes, along with the Summer and Winter Solstices, to help provide a rule of thumb for when things should be done. The Spring Equinox is a good time to get your seeds ready. The Summer Solstice is all about the sun getting higher in the sky, and this means better weather for your plants, so make sure they're outside now. When the weather starts to get colder and the sun starts getting lower, you will notice that your plants have started to fruit. But hold off on harvesting until the Fall Equinox. Finally, remember to have everything harvested and dried in time for the Winter Solstice.

This provides a useful window for when to start working on what step. Just remember that this guide needs to be slightly adjusted in order to fit your local area. A warmer climate doesn't need to worry about starting off indoors, they can go straight outdoors. Harvesting, on the other hand, has to be done earlier in colder climates. Use a growing calendar to get a starting time frame and then examine your local environment to fine tune your plan.

Growing Your Plants Indoors

When it comes to growing plants indoors, you have several options at hand for how you want to go about it. You could set up a tent in which to grow your plants, or you may even set aside an entire room to be your "grow room." Another popular way of growing indoors is to use what is called a grow box, which is just an enclosed system in which your plants grow. Each of these approaches can produce amazing results. However, regardless of the option you choose, there is still going to be a lot to set up when preparing to grow indoors. Understanding the eight steps necessary for growing indoors is the most effective way to cover everything you need and see if the process sounds like something you can handle.

The first step is to decide on the space that will be dedicated for the purposes of growing. As previously mentioned, this could be a closet, tent, grow box, room, or even just the corner of a room. When you are first starting out, it can be a good idea to start small. You are likely to make mistakes in the beginning and the smaller your setup, the less money is wasted on mistakes. You can also get a good sense of the growth cycle and how to care for plants best when you start small. Less plants means you can really take your time to watch and learn how they grow and evolve and see how often upkeep is demanded. However, keep in mind that you will likely want to expand your operation as time goes on. Picking a space which you can add more plants to is a great way to keep costs of expansion down.

You may have chosen to use a tent or a grow box for your designated area. These are smaller systems, so your way of tending to the plants will be to remove them from the growing area and work on them outside of their environment. You will need to have a space in which you can look after the plants while exposing them to the minimal amount of risk. Keep this in mind if you go with these smaller options; they do not take up a lot of space but you need to have some available for care. Tents and grow boxes are great because they are sealed against light, but you need to remember that your maintenance area has to be sealed as well. Your plants expect it to be dark at certain times and light at others. If light gets in when it is supposed to be dark, it may confuse the plants and detrimentally affect their growth.

Remember that your growing area is not just going to be filled with plants. You also need to have enough space for important equipment, such as lights. High quality lighting is the most important part of your operation. There are three kinds of lights that see the most use for cannabis growing.

High intensity discharge (HID) lights are the standard of the industry because they balance efficiency, value, and output. They will cost you more than the other types, but they shine bright as can be. You can get them in metal halide or high pressure sodium designs. Metal halide bulbs are great for vegetative growth while the high pressure sodium bulbs are most effective during the plant's flowering stage, and some growers opt for both designs to maximize efficiency.But if you can only afford one type of bulb, high pressure sodium bulbs are the better

ones to use. HID lights will also need a ballast and a hood or reflector for each bulb used.

Fluorescent grow lights are very popular for growers that see cannabis gardening as a hobby. High output bulbs are the most popular fluorescent grow lights because they provide a good amount of light but do not require a cooling system like HID lights tend to. They also cost less when building your setup because they usually have everything you would need, such as reflectors or ballasts, in a kit for purchase. However, fluorescent grow lights produce around 25% less light per watt of electricity than the HID bulbs do. This means that you will typically need to use more fluorescent grow lights and this could easily crowd your grow space.

LED grow lights have started to be used more often because of how efficient they are, but as a downside, they can also cost upwards of ten times the amount of a HID system. But these lights last for long periods and use less electricity. This means that they also produce less heat, which may save you on the cost of cooling. They offer the widest spectrum of light outside of the sun, which means better yields.

Regardless of what lighting source you decided on, your plants are going to need some ventilation. Your plants need both fresh air and CO2. You're going to want to have air flowing through the growing space for this reason. A typical setup for this is to have an exhaust fan near the top of your room with a filtered inlet close to the floor on the other side of the room. No matter how much air you have moving, you need to ensure that you stay within 70°F-85°F when the lights are on and 58°F-70°F when the lights are off. Because temperature is so important, you will likely find yourself matching your exhaust system to your plants' temperature needs. One way to figure this out is to turn your lights on for a few hours and then take the temperature. This allows you to see how much you have to control through your ventilation.

The next step, which most growers take but isn't a necessity, is to automate as many pieces of the system as possible. You can use timed switches to control the light/dark cycle or get a smart thermostat to monitor the space's temperature and control how the ventilation is running. This step is used to give you more free time, since most people cannot change their schedules to fit the needs of their plants.

Growing outdoors means that you typically rely on the growing medium provided for you. But when you grow indoors, you pick that medium ourselves. Soil is the most widely used medium because it allows the widest room for error. Another highly popular method is to use a hydroponic system, in which the plants take their nutrients from a liquid solution you have created to provide them with everything they need. A hydroponic setup is going to be much more expensive and detailed, but there are quite a few advantages that make hydroponics an impressive approach. While considering the medium, you should be considering what sort of pot or tray you will grow the plants in. Your medium may determine your container. For example, a deep water culture means your growing medium must be water.

No matter which medium you choose, the plants will want access to nutrients. Cannabis plants need nitrogen, phosphorus, and potassium, as well as calcium, magnesium, iron, and copper. Some growing mediums are fertilized for you. But if yours is not, then you will need to create a nutrient solution yourself. You can purchase nutrients in liquid or powder form and create your own water-based mix.

Finally, you are going to have to water your plants (unless you choose to use hydroponics). This means you will want to have easy access to your plants and have enough space for them to drain away. This takes up more space from your growing space. It is vitally important to have a proper drainage system in place, too, because stagnant water is unhealthy for your plants and can breed harmful viruses.

All together, growing indoors requires you to have enough space to properly tend your plants. A tent and grow box may allow you to keep the space small, but you still need to be able to take the plants out and tend for them properly, so leave yourself enough room to do so. Growing indoors comes with costs that the outdoors avoids but the control you have over your plants when growing indoors is unparalleled.

Chapter Summary

- Where you grow your cannabis is ultimately a choice between indoors and outdoors: control or nature.

- Growing indoors will let you take control of the process and have total power over the grow environment.

- Growing indoors also tends to produce better quality product but in much smaller yields.

- Both growing indoors and growing outdoors can be surprisingly expensive depending on the size of your operation.

- Growing outdoors is dependent on a seasonal calendar while growing indoors allows you to artificially make the conditions the cannabis needs to grow.

- Higher CO_2 levels when growing indoors allows the cannabis to grow faster and with higher THC levels.

- You can grow just about anywhere outdoors so long as the plants get enough sun and aren't exposed to overly strong winds.

- Growing near a river or lake is great for giving the plants the moisture they crave.

- When growing outside it is important to consider security as thieves are a real issue when farming cannabis.

- The growing calendar for outdoor crops can be divided into four sections: the Spring Equinox when you have your seeds ready; the Summer Solstice when your plants should be in the ground; the Fall Equinox for when you are harvesting; and the Winter Solstice by which everything should be dried.

- The first step to growing indoors is setting up a location that you can control, keep clean, have access to the plants and enough space for lights and equipment.

- HID lights are powerful but costly. Fluorescent grow lights are popular but create less light. LED grow lights are expensive but use less energy while also providing plenty of light.

In the next chapter, you will learn about cannabis seeds and how to differentiate between sativa, indica, ripe, and premature seeds. You will also learn how to germinate your seeds to prepare them for planting. Then, you will learn how to plant them and the lighting, temperature, and airflow your seeds need to grow, as well as how to care for your seedlings once they have sprouted.

CHAPTER FOUR

GERMINATION OF SEEDS

While growing marijuana focuses primarily on the bud that is produced, every plant begins with a seed and so shall we. In this chapter, we will contemplate how to pick our seeds, plant them, properly care for them, and see them through to the next stage in their growth. It is only from beginning with and properly nursing our seeds that we can achieve our desired results down the road.

Picking the Right Seeds for You

Cannabis seeds are only produced by female plants after pollination by a male plant. This is important to note because it tells us that the female plant is more valuable to us than the male plant because it can actually allow us to harvest our own seeds. We'll need to purchase or acquire seeds to begin with, but if we are attentive gardeners then we will be able to use our own seeds in the future. When we plant a regular seed, there is a 50/50 chance that the seed is male or female. However, only female plants produce the bud we want to harvest, so we are doubly interested in having more female plants. We can do this by purchasing seeds that have been feminized. This way we can ensure that we'll grow lots of female plants.

Just remember, we still want some male plants in order to fertilize our own females.

In order to feminize a seed, a seed is either sprayed with a solution of colloidal silver, treated with a method called rodelization, or sprayed with gibberellic acid. You can purchase seeds that have already gone through one of these processes and then use the process yourself with your own seeds down the road. These seeds will be almost identical to the female plant that parented them. This is otherwise known as cloning a plant, and we'll be looking at how that is done in the next two chapters.

When you are choosing which seeds to plant, there are a few factors that you can use to help you make your decision. The color of the seed is important. A healthy seed is dark in color, generally tan, brown, or black. Seeds that are lighter or even green should be avoided. Seeds will also have patterns on them that you can use to figure out if they are sativa or indica. Sativa seeds are typically a solid color while indica seeds have a striped pattern. You should also check how hard your seeds are. The harder the seed, the better it is. Do not even bother planting a soft seed; they are unlikely to produce a healthy plant.

Seeds come in various qualities, but the highest quality seeds are those that were harvested only after they were given a chance to fully mature. The environment the seeds are exposed to is also relevant, with the highest quality seeds being those that are stored in dark, dry places where they can cool off. If the seeds are allowed to mold, then they aren't

going to be worth planting. Keep in mind the age of the seeds. If they have not been frozen, then they should be no older than sixteen months. If they have been frozen, then age becomes irrelevant. What really makes high quality seeds stand out from the rest is their genetics. Great genetics means high quality plants. Ask about the genetics of the seeds before you buy them. Find out what strains the parents were to see if they are compatible and make for desirable buds.

Finally, while all the talk about quality and care is important, the real answer to what seeds you should pick comes down to what you are looking to grow. Do you want to grow a sativa or indica or even a hybrid? Pick the right seed for what you want to grow. Make your choice based on strain and use these practices to weed out lower quality seeds so that you only plant the best.

Planting the Seeds

What we want to do when planting our seeds is provide everything they need to germinate. This is the process through which a seed becomes a plant. Marijuana seeds need water, heat, and air in order to properly germinate. While there are many techniques used to germinate seeds these days, the easiest is still the most common. You will want to have a couple of clean plates and a roll of paper towels at hand.

Soak four sheets of the paper towel in distilled water. You want them to soak, but you do not want them to leak water. Drain away any excess water before you set them down on the plates. Put two sheets onto each plate. Next, take your seeds and place them on the plates with about an inch of room between them. Once they are set out, use the other two sheets to cover them. You should have two plates of seeds sitting on soaked paper towel with wet towels covering them. You want to keep these seeds in a dark place where they will be protected from harm. One way to do this is to use another plate, flipped upside to create a lid. Store the plates somewhere between 70°F-90°F.

You're going to have to wait for nature to do its part now. Check on the plates from time to time to make sure that the paper towels are still wet. If you find they are drying up, then adding some more water will be beneficial for the seeds. Rewet the towels by sprinkling water on

them or using a spray mister. Some seeds germinate quickly while others take some time, upwards of a few days even. You will know that a seed has germinated properly if it splits apart and a single sprout comes out. This sprout is called the tap root and it is the main body of the plant. As the plant grows, this root remains at its center. Keep in mind that you do not want to touch your seeds until they are ready to grow, as premature handling can cause issues in the splitting or the emergence of the tap root.

Once the seeds germinate, it is then time to plant them. Most growers transfer their seeds over to small pots initially. Doing this allows you to focus on the plants individually, making sure they grow strong enough to make it through to joining the main tray. Fill up your trays with a suitable potting soil, one that is breathable. Use a finger to make a small hole, about ¼ of an inch down. You will want to use tweezers or something similar to move the seed. It's important to be gentle now. Put the seed into the hole with the tap root down. Cover the hole and the seed back over with soil. Water the soil, starting with a spray bottle. You want to start with a spray to avoid drowning the seed. Keep an eye on the temperature of the pot to make sure it stays within the proper ranges below.

In five to seven days, you will notice a small seedling beginning to rise out of the pot. If you do not, there may be an issue with the temperature, the amount of water, or even the seed itself. Give it a couple more days before giving up on the seedling, as it may just be a late bloomer. Seeds all grow at different rates, some quick and some slow, and some just do not grow at all. Do not worry if you end up with a slow grower, it could still end up being one of your best harvests.

Lighting, Temperature, and Air

When you are dealing with seeds, it is important not to let them have any light. Once you have planted them in the soil, you still do not want to expose them to light yet. However, they still need to have the right temperature while they are growing into seedlings. The ideal condition for the seeds is between 68°F and 82°F. They also want a bedding that is moist, so you should be checking the soil at least once a day and giving it a spritz if it is too dry.

Turn on the lights once the seedlings start to emerge from the soil. The seedlings want to have at least 18 hours of light each day, so be prepared for your grow room to be illuminated pretty much at all times. In fact, you can actually allow the seedlings to have 24 hours of light at this point without harming their growth too much. They are going to need a more diverse lighting schedule once they grow past the seedling phase. Since you are unlikely to be there to watch this happen, it is a good idea to use an 18/6 hour light/dark setup, which mimics a natural day/night pattern.

You can use HPS or MH lights with your seedlings, but they are not needed. Seedlings do just fine under more cost-effective LED grow lights. However, if you already have another type of light, then you can use it. MH lighting is more effective for seedlings than HPS lighting, as seedlings prefer the cooler blue light found in MH bulbs compared to the warmer red light in HPS bulbs.

It should take between ten to fifteen days for your seedlings to reach the vegetative stage of their growth cycle. Of course, it is important to avoid overly high or low temperatures, either of which may result in the plants' stunted growth or deformation.

Though often overlooked at this stage, it is important to note that the seeds need to breathe. If they do not have sufficient access to oxygen, they will suffocate. This is one of the reasons you want to use a spray bottle to lightly mist the soil rather than pour water directly on the soil. If the soil is heavy or soggy, the seeds will be unable to breathe. When too wet, the soil becomes a suffocating pillow over the seeds. The seeds need both moisture and oxygen.

You should also be careful not to plant seeds too deeply. When seeds are too deep, the soil above them becomes a suffocating weight, and even if the seeds do not suffocate from the soil, they face an entirely different issue. The shell of the seed provides needed energy to the seedlings to help them dig into the soil and set their tap root. But if the seeds are buried too deep then the seeds will not have enough energy to break out of the soil. You will have a bunch of popped seeds suffocating and dying before they can even reach the top. Always keep the seeds a quarter inch deep, do not go any deeper.

Caring for Your Seedlings

We've seen how our seeds need to mature, can be kept for sixteen months and more when frozen, and like the dark. Taking care of your seeds is mostly done by leaving them alone to look after themselves. But your seedlings, on the other hand, are going to need some tender loving care. Cannabis plants are most vulnerable during the seedling stage, as they are very small and easily stressed or damaged.

Seedlings are at risk of being eaten, damaged by the weather, drowned, or worse. They do not have any real strength yet, and so it is up to you to ensure they're protected. One way to protect them is to use a transparent plastic bottle. Cut the bottle in half and stick it in the soil, so that the seedling is in the middle of it. This setup protects the seedling

from pests while still allowing light to shine through. Make sure you cut air slits into the bottle, as the seedlings still need oxygen.

You want to make sure that the soil is moist, but overwatering can be a costly mistake. Too much water, as previously touched on, can suffocate the plant. We start our seeds in small containers rather than large ones because there is less risk associated with suffocation or waiting. In addition, the roots of the seedling do not sufficiently fill a large container and hence cannot reach all the water and nutrients it needs. But plants in a small container are easily overwatered and end up damaged for it. This problem changes as the seedling continues to grow. It will eventually get to a point where it's root system is too large for the small container. When this happens, you will have to move it to a larger container, or the root system will tangle and trap water in the container.

But just as overwatering can be a problem, underwatering your seedlings is also a problem. The roots of the seedling need to have access to plenty of water at all times. A lack of water will result in the plant wilting and the leaves staying too small or frail. If the soil in the pot is separated from the walls of the container then the environment is too dry for the seedling and you need to water it immediately. Plants that are grown in enriched soil risk growing strangely when they haven't been watered. They take on a dark shade of green and start developing growths in the wrong places. If you notice this, then you need to water the plants properly to save them. Check your plants every day to ensure that they have enough water.

Your seedlings will want to be fed nutrients. But you need to be careful not to create a toxic environment through overfeeding. When soil has too many nutrients in it, we call it "hot soil." This is because the plants will start to appear as if they have been burned. The tips of the leaves will be discolored, and you will need to water them with distilled water, which has had no extra nutrients or minerals added to it. But even if you have faced a toxic environment, they are still going to want to be fed. Now, however, you will need to be particularly wary about the nutrients you give to the plants. A good rule of thumb is to begin giving the seedlings a half dose compared to what you had been giving them prior to the hot soil issue. You can continue to increase the nutrient dose back to the package recommended dosage so long as you watch for issues.

The flipside of the hot soil issue is a lack of nutrients. If your plants are starving to death then you will notice that the leaves begin to grow all folded up. If you see folded leaves then you are going to want to add some nutrients to their mix. You may also want to check the pH level of the soil to ensure the soil is in a healthy range while growing. If the soil pH is outside of the optimal growing conditions, then the problem is that the nutrients in the soil aren't being absorbed properly. If you do not water them properly then they also will not be able to get the nutrients they need.

If you follow these steps and take care of your seedlings then you are well on your way towards your first harvest.

Chapter Summary

- Only the female plant produces seeds and only then after being pollinated by a male plant.

- The female plant is also the only one that produces bud, and so it is by far the most valuable of the plants.

- You can feminize cannabis seeds so that they sprout as females, where planting regular seeds gives you a 50/50 chance of getting a male or female plant.

- A healthy seed should be dark in color. Light colored seeds are premature. A healthy seed is also hard while a premature one is soft.

- Higher quality seeds are stored in dark and dry places.

- We need to first germinate our seeds, which can be done simply with a wet paper towel press.

- You will know that a seed is ready to move into the soil when the shell cracks and a tap root has emerged.

- Seeds should only be planted ¼ of an inch into the soil and then sprayed with water.

- While in the seed phase, these plants want no light.

- After five to seven days in the pot, a seedling should be seen emerging from the soil.

- Seedlings like having 18 hours of light, a temperature between 68°F and 82°F and plenty of airflow.

- Overwatering seeds can kill them because the soil becomes too heavy and tight for oxygen to get through.

- Seedlings are small plants and can easily be harmed. Be very careful when doing anything with them.

- Underwatering your seedlings can cause they to wilt and look frail.

- Seedlings will be started on a nutrient diet but it is important not to overfeed them and create a toxin environment. If this happens then the environment will need to be rinsed with distilled water.

- An environment without enough nutrients will also harm seedlings. You can tell this is happening if the leaves of the seedling come in folded up.

In the next chapter, you will learn what a mother plant is and why it is so important when it comes to cloning plants. You will learn everything you need to know about picking the mother plant that is right for you and how to grow one that you can use again and again.

CHAPTER FIVE

WHAT IS A MOTHER PLANT?

When you find a strain that you absolutely love, you will want to use it repeatedly. The best way to make this happen is to keep around a mother plant, which is a plant that you keep in order to clone. Mother plants are kept in the vegetative state and aren't allowed to move into the flowering stage. Having a mother plant allows you to speed up the time needed for cultivation because you can simply clone the plant instead of repeatedly growing a new one from seed.

Keeping a mother plant on hand is highly efficient because it can reduce the time between cloning and harvesting to mere days. Choosing to clone instead of growing a daughter plant also minimizes any crop variations by ensuring that the characteristics of the strain stay consistent between plants and that every clone is a female. In this chapter, we'll see how to choose, grow, and maintain a healthy mother plant. The chapter after that will focus on how we use mother plants to clone plants.

Choosing a Mother Plant

Mother plants are important to growers. Your mother plant is the source of much of the marijuana you grow, and it is the method through which you ensure that you are producing high quality buds on a consistent basis. Your mother plant is special, the creme of the crop. So it is important that you pick the perfect plant. The plant you pick is going to depend on what you enjoy. If you enjoy indica-type strains, then you aren't going to be growing a sativa mother plant. But regardless of

personal preference, there are some tricks that you can use to ensure your mother plant is among the best of the best.

Nowadays there are many styles of feminized and autoflowering seeds available on the market. With feminized seeds you know you are guaranteed a female plant, which seems like a good fit for your mother plant. But most growers will tell you that you should only use normal seeds for growing your mother plant. Feminized plants have a higher risk of hermaphroditism. Your mother plant will be trimmed a lot to make clones, and this practice stresses the plant. When a feminized plant undergoes a lot of stress, there is an increased likelihood that the plant will flip its gender and start to grow male flowers. If such a phenomenon occurs, you will have to start all over. It is better to plant regular seeds and find a female the traditional way.

It is generally agreed that working with F1 hybrids is the best way to go about picking mother plants. An F1 hybrid is the first generation child of a cross between parents of two different strains. Your mother plant should be (though it doesn't have to be) one that you bred yourself by crossing strains. F1 plants tend to have the best aspects of their parents, and this means they have great genetics. F1 plants are also pretty quick growing compared to other seedlings, and this is great for maximizing your yield:growth ratio. You may also want to use an F2 hybrid, which comes from crossing two F1 hybrids. These plants are also amazingly potent, but on the downside, they take a lot of time to breed because you need to grow four plants, breed them twice, then grow two plants and breed them once. A single F2 hybrid requires six plants and three crosses. You can always work with regular seeds, but F1 or F2 plants will be the most impressive ones to turn into mothers.

Since we use regular seeds rather than feminized ones, we will have to take the time to germinate them first. Grow the seeds as you would any other cannabis seeds until they reach the vegetative state. Then, starting three weeks into the vegetative stage, take some cuttings from them. This should be done early enough that the plants are still pre-flowering but late enough that the cuttings can take root. You want to take pre-flowering cuttings because the clones will be the same age as the cuttings. If your plants have already flowered, then your cuttings will as well. Take some time and label your cuttings so that you know which plants they came from. You will want to keep growing these cuttings like you did any normal plant, just remember to keep them labeled and to

keep them separate from the original ones. Once your original plants flower, take note of their gender and remove any males as soon as you spot them. Head over to the clones you were growing (the cuttings) and remove the males from this group as well. This step is done to prevent them from pollinating a female before you are ready for such.

With the original plants flowering, keep your eyes on how they grow. If you are working with hybrids, then you will especially notice the way that each plant has its own unique features depending on which genes it received from its parents. You want to ensure that you turn the best plant into the mother plant. Which plant is the best will be up to you, but there are some markers that can help you decide.

Look at how vigorous the plants are. Since your mother plant will be trimmed often, you need a vigorous plant able to withstand the stress of cutting. Another trait that growers look for is the physical appearance of the plant. Some might prefer to smaller plants to save room, while others are partial to bigger plants. Coloring, spacing, and leaf size are all features that you may consider importance as well, depending on your taste. Yield is also a determining factor. You may want huge harvests, regardless of quality, or you may want smaller harvests of high quality. It's up to you. Also, use your nose and smell the plants. This may be a deciding factor in which you pick as your mother plant.

Growing a Mother Plant

Start with seeds and grow them through until they are in the vegetative growth stage and you can see what sex they are. Identify all of the females and take clones from them. You are going to want to take several clones from each female and then label which they came from for later reference. Continue to grow the new clones through to the flowering stage. You are going to keep the original plants in the vegetative stage. Remember that marijuana plants naturally react to the changing of the seasons and begin to flower when the days get shorter. To keep it in the vegetative growth stage is as simple as leaving the lights on it for 18 hours a day. Doing so tricks the plant into thinking it still isn't time to flower yet.

Your newly cloned plants will travel through to harvesting, and you should take the time to really examine how the yield turned out, the smell, the flavor, and everything else. When you have found the best plant, use that for your mother plant. Determine the mother of your preferred clone, and use that one as the overall mother plant.

A lot of growers like to offer extra protection to their mother plants by first growing them in an organic base. This is done so that the mother plants have the immunity necessary to fight off disease and stay strong. You will also want to be sure to use the original plant as your mother plant and not the clone that came from it. Plants that have been grown from a seed will have much stronger roots and immune systems.

It is worth considering the use of a nutrient solution specifically designed for mother plants. These mixes have nutrients to keep strong clones and healthy mothers. You really want your mother plant to have strong cell walls and high carbohydrate levels. Using too many nitrogen-rich nutrients is not good for mother plants, as doing so can lead to thin cell walls and low carbohydrates. Nutrients with a lot of calcium will help to improve both of these.

Mother plants will eventually start offering worse and worse clones. They'll grow much less vigilantly, while producing worse marijuana. If you are careful and take care of your mother plant, you can get a couple years out of it. But when the plant starts to offer diminishing returns, then you know that it is time to start considering what your next mother plant will be.

Chapter Summary

- A mother plant is a plant that is kept in the vegetative state to be used for cloning purposes.

- Your mother plant will likely be the best strain you can grow.

- You can use feminized seeds to get a mother plant but it is better to grow your mother plant from natural seeds instead.

- F1 hybrids are widely considered to be the best plants for use as mother plants.

- When your mother plant enters the vegative period you will take cuttings from it and grow these as cloned plants.

- You want your mother plant to be vigorous so that it can withstand the stress of all the cuttings you take from it.

- Take several clones from each potential mother plant you are growing so you can see multiple examples of how well it clones.

- Newly cloned plants will take less time to reach harvesting than those you grow from seed.

- Growers like to use an organic base when raising a mother plant so they gain immunities needed to stay strong.

- Nutrient solutions are also powerful tools in helping mother plants stay healthy.

- Mother plants will eventually start to create weak clones, remember they do not last forever.

In the next chapter, we will continue our discussion of mother plants by looking at how you go about cloning your plants. We will cover the benefits of cloning plants to better understand why it is widely popular with both professional and hobby growers alike. You will then learn how to clone your own plants, which involves making more mother plants as well. We will also go over some of the mistakes that

new growers make when they first start out, so that you do not fall into the same traps that others have.

CHAPTER SIX

CLONING YOUR PLANTS

We talked about cloning a lot in the last chapter because we actually have to first clone several potential plants to decide which should be our mother plant. It's easy to see how mother plants are intimately involved in the cloning process, but what we haven't touched on yet is the many reasons why you would want to clone a plant in the first place as opposed to just germinating and growing from a seed. First we'll tackle why you would want to clone a plant, then follow this with a detailed look at how you clone a plant. Combined with Chapter 5, this chapter will equip you with all the knowledge you need to choose your mother plant and clone your first plants.

Benefits of Cloning Cannabis Plants

We spoke earlier about how you need to fertilize a female plant with a male plant in order to create seeds. These seeds then allow for the growing of new plants and this new plant will have features from both of its parents. This is the most classical form of breeding that there is. But it there is another newer way to breed a new plant through cloning. Cloning is a form of asexual reproduction because it does not require a male plant. To clone a plant is to take a cutting from it that will then grow into a new cannabis plant that is identical on a genetic level to the plant it was taken from.

Cannabis cultivation via cloning has taken off both on the industrial level as well as at the home grower level. This is because cloning offers many benefits that cannot be found through traditional breeding practices. Cloning a plant can save money, as you already have the materials necessary, and you do not need to go through the seedling phase that eats up so much energy. But that is only the most basic of the benefits that cloning offers.

If you grow an amazing plant that has high quality buds, you naturally want to grow more of that kind. Traditional breeding would mean that you have to combine two plants's genetics together. This would give you a third plant with genetics taken from its parents. It might produce buds of a similarly high quality, or it may grow buds of a lesser quality. That is the inherent risk of traditional breeding. But if you clone your plants, then you end up with an exact replica of your original plants with the same high quality buds. This means that you can keep growing the best stuff for quite some time to come.

Cloning is also useful if you want to isolate and replicate particular features. If you have a plant that grows quickly, then cloning it will allow you to produce many quick-growing plants. Or, if you have a plant with a flavor you really enjoy, then you can clone a bunch of these. If you have a higher number of plants on hand with your ideal features, you can experiment more and mate them the traditional way and see what kinds of plants you can come up with. This makes cloning particularly great for those who really enjoy getting deep with their crops.

Another great feature is the fact that cloning can basically give you a garden that is never empty. By skipping the earlier stages of the plant, you are able to shorten the time it takes to go from seed to harvest by only going from vegetation to harvest. While you will find that your clones get worse over time and you will need to replace your mother plant, you can get a lot of mileage out of cloning your plants. What's even better is that cloning isn't very hard at all.

How to Clone a Plant

First things first, you are going to need to grab a few things before cloning your plants. You will want to get something to cut the plants

with. While many people reach for scissors, they actually aren't a great choice. Scissors often crush the branches they are trying to cut through, which can make it harder for the plant to take root. Instead, a razor that cuts but doesn't crush is a more professional choice. You will also want some water and the rooting medium and hormone that you will be using.

The best plants to clone are the mothers that are healthy and sturdy. They should be a couple months into the vegetative cycle as well. You can clone a plant as early as the third week of the vegetative period, but the clone will have a much better shot at rooting properly if you wait a little later.

You shouldn't just grab your razor blade and start hacking away at the mother plant. There are actually some steps you will want to take in the days before cutting. These steps will help to ensure that the mother plant receives as little shock as possible so that it can stay healthy and continue giving you clones for many harvests yet.

The first step you will want to take is to stop fertilizing the plant ahead of the cutting. The fertilizer has nitrogen in it, and while the plant needs nitrogen, it is better to allow time for the excess nitrogen to work its way out of the leaves before cutting. The reason you do this is to make it easier for the clones to take root. Excess nitrogen in the leaves and stem of the plant would signal the clones to try to grow vegetation when you want it instead to grow a root.

Remember to keep the work area sterile. This is where it becomes extremely important to have a proper location set aside for working on the plants if you are using a tent or grow box as opposed to a grow room. The environment should be sterile so that the freshly wounded mother plants and the cuttings you take do not become infected. A sick plant isn't a happy plant, after all.

It is time to start cutting. But first, you should determine the branches you want to cut. It is best to start at the lower branches which are generally the sturdiest and healthiest. You should be taking cuttings that are about 8-10 inches long. These cuttings should also each have several nodes on them. Once you have picked out the sections you want to cut, you will want to cut as closely to the main stem as you can. Your razor should be held at about a 45 degree angle to the branch. This is

done to increase the surface area of the rooting space. A larger surface area means faster growth.

Now you should be holding a cutting in your hand. You want to immediately throw that into the water that you prepared at the beginning of the process. The water will prevent air bubbles from forming in the stem. These bubbles are deadly because they prevent the stem from absorbing water. An air bubble in the stem of your cannabis plant is pretty much as deadly as an air bubble in your bloodstream, and it can quickly kill a cutting. There are a lot of growers who cut along the stem of the cuttings even more at this stage as it purportedly helps to increase the rooting potential of the clone. It is a good idea to take the time to do this yourself so long as you do it prior to putting the cutting into the water.

With several clippings taken from your mother plants and moved over to water to keep safe, it is a good idea to clip their leaves. This helps the new clones have a more sanitary environment for when they root. It also helps the clones to make use of the processes of photosynthesis in a healthier manner. Use your razor, or even scissors on this part, and clip the fan leaves from the top down to about halfway down the stem. If there are any leaves near the bottom of the stem then you can remove these entirely so that none are left to just lay against the growing medium. Taking care of the leaves at this stage helps the clones to make better use of water and nutrients while rooting.

If you are working with rooting hormone, then this is a good time to drop the stems into your solution. Rooting hormone comes in gels, powders, and liquid solutions that have been formulated to help promote rooting and growth. You only need to dip the stem into the solution for a moment before you plant the clones into their new home. That home should either be Rockwool cubes, which allow for excellent airflow and hold moisture in well, soil, or straight water, if you are running a hydroponic setup. Each of these options have their strengths and weaknesses, so choose the one that is most appropriate for you and your grow space. You will be lighting these clones as if they were seedlings and providing at least 18 hours of light. You can purchase automatic tools, called auto-cloners, which take care of the water, oxygen, and light needs of the clones but these are quite expensive. For

your first clones, I recommend taking care of the plant's needs yourself so you can get a proper feel for what goes into cloning.

The final step in cloning is the transfer back into the garden's circulation pots. You can tell that a clone is ready to be moved when you start to spot new vegetative growth beyond what it was planted with. Remember to be careful in moving the plant as it is still weak and you do not want to give it transplant shock, a common problem with cloning that arises due to overexertion of the plant and poor sanitation prior to transplantation.

Now you can sit back, watch your clones grow, and prepare for harvest. Once you have your seedling trays empty again, you can go ahead and start looking towards the next clones you want to prepare. Your mother plants will stay lit and stuck in the vegetative stage while your clones grow up, harvest, and earn you income. It should be a decent haul if you stick with cloning your most impressive strains!

Mistakes New Growers Make When Cloning

We'll be looking at more mistakes that new growers make when it comes to cultivating marijuana, but here we are going to speak specifically about the most common mistakes that happen during the cloning phase. These simple tricks can help you to avoid a lot of pain and wasted time.

We've talked about it before but it has to be mentioned again: You must, absolutely must, keep your working area clean and sanitary. Use isopropyl alcohol to wipe down all of the surfaces that you will be working on. It is even a good idea to wipe down the surfaces around where you are working just in case you set something on them without thinking about it. A single mold spore can rain havoc on all of your

cuttings. It is also a good idea to get single-use razor blades so that you never use an old razor for a new cut. Cut everything you need to for this batch of clones then dispose of the razor and use a new one next time you are ready to do some cloning.

There are all sorts of rooting hormones out there on the market these days. But not all of them are worth their price. Many of them are downright awful. It is always a good idea to look at the reviews and see what people are saying about them. One of the best out there is Dip-n-Gro, which is used by professional arborists and cannabis cultivators the world over. Speaking of roots, there is a ton of writing on what is the best temperature for your roots. They have found that the best temperature to speed along new root growth is 78°F. There are many ways to achieve this heat, but one easy answer is to use a seedling heating mat. Just remember to get a thermostat that can be stuck into the growing medium. While we're talking mediums, you should use the same medium for your seedlings that you do your main crop.

Some people seem to think that clones need 100% humidity. There are actually some downsides to this. When you leave the humidity up that high you are allowing mold and wilt a window in which to grow. One idea to get around this is to use a humidity dome on your new clones for a couple days but make sure to get ones that have holes in them. The holes allow vapor to pass through and leave, which brings the humidity down lower while still providing help to your clones. The best part is that those vents make it nearly impossible for those pesky pathogens to grow.

If you are using a canopy, then you should consider the temperature it is at. 86°F is great for maximizing carbon dioxide uptake. However, it can be hard to keep this temperature since the rooting zone needs to be around 78°F. If you can get the canopy anywhere between those two numbers then that's awesome. The root zone temperature is more important than the canopy temperature but that doesn't mean you should just completely ignore the latter.

You can take some serious control of your lighting to really help out your new little clones. They need at least 18 hours of light, so plan to give them 18 hours of light and 6 hours of darkness. The best lighting for helping the roots to take hold is 120-200 umol m-2 s-1. If you are using T8 or T5 bulbs, then this shouldn't be any problem. As the roots

start to take hold and grow, you can turn up the intensity by simply bringing the light closer to the plants and double checking the intensity. It can be useful to get yourself a PAR meter to be able to measure the light. In fact, you should just get one in general because there are all sorts of times that a PAR meter earns its weight in gold when dealing with your indoor cannabis operation.

Another mistake that is often made is to start your cloning with clones. The problem here is that when you use another clone that isn't quite mature, you are put in a spot where you will not really be able to predict how the plant's yield will turn out quantity or quality wise. If that plant turns out to be of poor quality then so will the rest of the clones you have growing. It is always better to start from a seed and work your way out from there. That way, you know what you are getting involved with.

Many new growers will set themselves up with a mother plant that they absolutely love. It produces high-quality product or has the traits that they enjoy the most. So of course they will clone it like crazy. But the more this happens, the more they'll notice that later generations of clones aren't of the same quality they fell in love with. The THC and CBD levels degrade in time, and same with other elements such as smell, color, and sturdiness. Pay attention to the health of your mother plant and prepare to change it regularly, depending on how often you are cloning it. Most professionals advise changing mother plants after the fourth generation of clones.

A major issue is forgetting the water when cutting. You should either have water nearby to put the trimmings in or you should make your root cuts under running water. Either of these options work, the key is having that water to destroy any air bubbles in place. The problem with forgetting the water is that you will not immediately realize that an air bubble is killing your plant. The plant will get sick and slowly die, rather than just die off immediately. Because of this, you may find yourself checking the soil and nutrient solution when the problem was simply that you forgot to clear the trimmings before potting them.

Hopefully these simple tricks give you the advantage you need to be cloning amazing plants in no time.

Chapter Summary

- To get seeds you need to have a male plant fertilize a female plant. This means that seeds will always require the genetics of two plants and that means a variety in the genetics.

- A cloned plant will always have the same genetics as the mother plant it came from and this means an identical plant, one that you know exactly what it provides.

- The predictability of cloning means the practice is very popular at all levels of the industry.

- Cloning can save money and allow growers to get more yield and income off of their best strains.

- The seeding phase of growing a plant uses a lot of energy. Cloning a plant can help you to save money by skipping this phase entirely.

- Cloning is great for isolating and replicating features that you want to continue breeding into further plants.

- Since cloning a plant can be done any time of the year, this can leave you with a garden that always seems ready to harvest.

- To clone a plant you want to select healthy mother plants that can survive the cutting process.

- Always stop fertilizing your mother plants before you take cuttings, as this prevents excess nitrogen from slowing down the clones' rooting process.

- It is important to keep the work area sterile to avoid infection. Always wipe and clean anywhere you plan to work and the surrounding areas that you might accidentally touch.

- You want your cuttings to come from the lower branches that are around 8 to 10 inches from the bottom, and these cuttings should have multiple nodes on them.

- Use your razor to cut at a 45 degree angle on the branch so that there is a larger rooting space for your clones.

- Next you are going to want to put the cuttings into water to get rid of air bubbles. Air bubbles left in a clone will kill it slowly and leave you wondering what the problem was.

- Clip the leaves of the cuttings so that they can more easily use photosynthesis.

- Drop the cuttings into rooting hormone. You will also want to cut at the roots a little to create a more direct connection between the root and the soil.

- Plant the cuttings and you got yourself a clone.

- New growers often forget to keep the work area sanitary and this leads to deadly results.

- Experiment with rooting hormones to find the one that works best for you. Cheap hormones get poor results.

- Too high of humidity risks mold and wilt, clones should have high humidity but be allowed to breathe.

- Keep the temperature of the rooting zone around 78°F.

- The new clones will want at least 18 hours of light.

- Do not start cloning from clones, always clone from a plant grown out of seed.

- Remember that mother plants degrade, plan to replace it after three generations.

- Never forget the water when cutting, it is necessary to keep your plants alive and well after all.

In the next chapter, you will learn how to finally harvest your crop. You will first learn how to tell that your marijuana is ready for harvest by examining the color of the trichomes. Then you will learn a step-by-step process for how to harvest your plants yourself. This guide will go over how to prepare your workspace, flush your plants to prevent

contamination, physically harvest the plants, and then trim them to look beautiful and appeal to buyers.

CHAPTER SEVEN

HARVESTING YOUR CROP

For many, harvesting is the most exciting part of the whole process. Who can blame them? If you consume marijuana then this is what you were building up for and now you get to enjoy the fruits of your labors. And if you are growing for money, well, now you get to make that income.

If you are growing outdoors then it has taken a long time, too. You've had to wait as it matured and grew. But is it fully grown, just because it is within that window specified in Chapter 3? Before you go ahead and harvest, it is always a good idea to take a deeper look at your plants to make sure they are ready. We'll see in this chapter how to determine the readiness of our crop, as well as harvest our cannabis, and learn about trimming it.

How to Tell If Your Marijuana Is Ready for Harvest

The window of harvesting is going to be set by the strain type rather than by the calendar we saw earlier. The calendar is a good guideline for the basic steps of when the harvesting opportunity is available, but it is always best to check your plants and respond to what they are telling you. While they do not have mouths from which to tell us how they are feeling, the cannabis plant has many ways of announcing that it is ready to harvest. This is important because when you harvest is more important than you may think.

Marijuana is primarily used to get high, and that makes the effects of the plant very important in regards to determining its value. The high will first be affected by the strain, and as you already know, different strains have different effects. Sativa strains have different effects from indica ones, and then even within a type there is a lot of range. Each strain has additional variety to it, as told by descriptors such as bubba, skunk, kush, haze, and purple. All of these words (and more) are used to describe a variation of the strain itself, and you will see these variations pop up across many strains, all to let the purchaser know what to expect. Finally, the most important in terms of harvesting is the amount of time that has gone by between the start of the flowering stage and the harvest. The amount of THC glands that have been allowed to develop is determined by how much time the plant has had to flower. The longer the time it has had, the more mature the plant is considered to be and the more THC it will have.

When you look at a marijuana bud up close you see leafy plant matter, little hairs, and numerous specks of white crystal. It is these crystals that hold all of the THC. In the cannabis community, this part is simply referred to as the crystal. However, they are really called trichomes. It takes a few weeks of flowering for trichomes to form, but they quickly start to multiple, grow, and mature once they initially form. These trichomes tell us how much THC the marijuana will have, and they also serve as the best guideline for deciding if your plants are ready to harvest because they help us estimate the plant's maturity at a glance through their color.

There are four trichome colors to understand. The first is translucent, in which the trichomes appear like little water droplets on the bud. This tells us that the plants are not fully matured yet. Essentially, these plants have very little THC, and nobody is going to purchase or use this marijuana if you harvested it now. Such a plant would also have a much smaller yield, since much of it would be unusable. On the other hand, if the trichomes are amber, then the plants are overly ripe. These plants, however, are still sellable, as overripe trichomes are linked to a "body high."

The majority of growers harvest their plants when the trichomes are a cloudy color. They still look like little water droplets on the bud, but you cannot see through them anymore. At this stage, the bud is at its ripest and most potent stage. However, some people like to wait just a little bit past this stage. As the trichomes start to become overripe and just take on an amber color, you can harvest them to get a combined effect in which the high is shared between the mind and the body. This makes for a widely enjoyed type of high that many people think is the best.

Indica strains can be expected to mature far quicker than sativa ones. Between six to eight weeks after flowering, you should be harvesting. Start examining the trichomes around week five to find the best window in which to harvest. Meanwhile, sativa strains will take between eight and twelve weeks to mature. In fact, sativa plants may take even more depending on the particular strain. To be safe, you should start checking the trichomes around week eight of flowering.

The best way to check on the trichomes is to put a bud underneath the microscope. Most photos of marijuana made the trichomes look huge, but they are really quite small and hard to see with the naked eye. Pretty much any pocket scope will give you enough magnification to see them clearly. You should expect the trichomes to be clear when you first look, then you want to keep checking them daily to watch as they shift. It is up to you whether you want to harvest for a head high, a body high, or a combined head and body high. It all depends on your goal with that particular yield. Remember to wear gloves and handle the plants carefully. You would hate to damage them now when they're this close to ready.

There are a couple other indicators that let you know it is harvest time. The trichome is merely the most popular and widely used method. You can also look at the color of the little hairs on the bud. These are called pistils or stigmas. Immature plants will have yellow-white pistils. By the time the plant is mature, they will be orange. There are also the swollen calyxes, which are little pods on the bud. These indicate if a plant has been pollinated, as they will have seeds in them if so. But in unpollinated plants, the swollen calyxes slowly swells throughout the flowering period as well. When they are nice, thick, and plump, the plant is signalling you that it is ready to harvest. You can also determine how mature the plant is by the overall size of the buds. They will continue to grow until they begin to slow and eventually stop. Growing buds are not the best sign for judging maturity, but it certainly can be considered when taken along side the other pieces we've looked at.

So if your plants are nice and mature, it's time to harvest!

A Step-By-Step Guide to Harvesting Your Plants: Preparation

The first step, before you pull the first bud off the plant, is to get everything prepared and ready for the harvest itself. You are going to want to set aside enough space for all of your plants. It is a good idea to handle them all in the same place, because otherwise the odor can get everywhere. Marijuana has a strong smell, and when you have an entire crop's worth, it gets especially smelly.

But you also want to set aside a designated location so that you can control the temperature of the room. You're going to want to keep it at 70°F or below. This is to stop the cannabis oils from volatilizing, which again will help to reduce the odor. If the room is too hot, then they are going to really start stinking up the place. It can be a good idea to seal the room and then install a vent, but this is only for really large harvests.

You're going to need some tools to handle this. First off, gloves are going to be a must. You will want them to be latex or nitrile. These will help to stop the buds from getting contaminated. They will also help to keep your hands clean when handling the fresh buds as the buds can be extremely sticky. You will also want to get trimmers, which could be scissors or electric trimmers designed for the job. Whichever you go with, you will be using them a lot, so a trimmer tray is also a good idea. Trays make it easier to sort the harvest and excess product, crystal or kief, can be caught in the trays so you do not lose anything when working through it. You will probably also want to set up some lines throughout the room that you can hang the bud from. These will help you to dry out the product.

With the tools prepared, it is time to prepare the plant next. The first step is to remove old leaves. These are the leaves lowest to the ground. You will notice that they have turned, or are starting to turn, yellow rather than green like the rest of the plant. These leaves are stealing energy away from the buds. However, it is important to only trim when you are preparing for the harvest. It is also important to keep your trimmers to the lowest leaves. Do not remove anything above the bottom quarter.

It is important that you stop spraying the plants. Whatever you are using on them, be it nutrients or some other specialized chemical process, needs to stop two weeks or more before you begin the harvest. These sprays will leave a residue on the buds, and nobody wants that. Stop spraying two weeks ahead of time to prevent this from happening, and decrease the chance of mold forming on the buds. Now it's time to flush the plants.

A Step-By-Step Guide to Harvesting Your Plants: Flushing

Flushing your plants is such an important part of the harvesting process that it deserves to be singled out. Flushing is done to remove any residual traces of the chemical or nutrient sprays that you used on the plants as they were growing. One of the reasons that people use marijuana these days and a key reason it is growing in popularity is the fact that it is a natural product and not chemically produced in a lab. The last thing anybody wants is to consume chemicals they didn't sign up for. It wouldn't be enjoyable for them, and it wouldn't look good for you or your product. Flushing is the easiest way to tackle this problem before it ever goes anywhere.

You might think that nobody would notice if there was residual traces of chemicals on your marijuana, but to a person experienced in consuming cannabis, it is as clear as can be. The taste of the smoke is different than it should be, harsher. Smoke with chemical traces on it burns the throat and lungs more and has a distant chemical smell. If you are growing organically, you might be able to get away without flushing, but if you are using any kind of synthetic fertilizer at all, then you absolutely must flush your plants. I recommend that you flush even if you are going straight organic because it is better to be safe than to be sorry.

Flushing your plants is going to take anywhere between a week to a week and a half. Two weeks before your planned harvest, stop spraying your plants and begin flushing them. Doing so forces the plant to use up excess nutrient it has already stored instead of taking up more from the soil. To flush, your plants simply start spraying them down, preferably with distilled water. If you have to use tap water then you can purchase a flushing agent to mix in to create your spray. If you are growing in soil

(indoor or outdoor) then you spray them down with clean water just as you would spray them down with the earlier treatments. If you are running a hydroponic setup, then you just switch out the reservoir with a clean water reservoir and let it circulate for three days. After three days, empty the reservoir and fill it back up with clean water again. Do this two or three time.

If you are flushing soil then you want to have about 15% run off, give or take 5%. You should also let the plants remain dry for a day or two before you harvest as doing so will help with the drying process. If you are using a hydroponic setup then you want to take out your PPM or EC meter and check the concentration of the reservoir when flushing. If the counts are close to what they were when you had the nutrient mix then you are going to want to keep emptying and replacing the reservoir with clear water until the numbers come back down.

Doing this takes time and effort but it ensures that you do not sell or consume tainted marijuana.

A Step-By-Step Guide to Harvesting Your Plants: Harvesting

It's time to actually start harvesting and bringing in the plants finally. Double check your environment. You want to avoid strong lights, temperatures above 80°F, friction from how you handle the plants, and overly damp or humid conditions. These are all harmful to the THC concentration. As the THC is the whole reason many people are looking to buy your crop, you want to make sure not to damage it this close to the finish line.

There are two ways that you can cut the plants to begin the actual harvest. The first way to to harvest the whole plant. This is the quicker of the two methods. It is also the easiest because it doesn't require that you analyze the buds beforehand. Just take the main stem and cut it off. You then take the whole plant and hang it up on your lines, or alternatively, you can cut it into smaller pieces and hang these.

The other way is to harvest the ripest buds only. This is a longer process because you need to examine all the buds in the process. Taking what you learned about the maturity of the buds, examine your plants and remove only those that are in the proper window of effect. Then, about a week later, return to the plants to harvest the buds that have matured since the last cut. The rule of thumb here is to start with the buds on the outside, those that have been exposed to the most light, as these mature faster than the ones in the shade. Doing this method takes a lot longer and requires you to examine and get to know the buds, but it can also increase the quality and size of your yield.

Put on your work gloves, they're not coming off until the rest of this is done.

Regardless of which method you use, you will find yourself looking to cut through the main stalk at one point or another. This is where your cutting tool comes in. The big fan leaves, which have no trichomes, can be removed from the plant. You might want to do this in order to make it easier for the plants to dry. If your workspace or drying space has high humidity then getting rid of the leaves is a good idea to help prevent mold from taking hold. A low humidity environment might be best if you want to leave the leaves on the drying plant.

After cutting the plants, hanging them will make it easier for them to dry in an even manner. In addition to checking the humidity, you want to make sure the area where the plants are hanging has good ventilation. Setting up fans or air conditioners can help ensure good airflow in your space.

A Step-By-Step Guide to Harvesting Your Plants: Trimming

To harvest your plants, you cut them down and hung them up. Now it's time to give your plants a manicure, or trim, so you can see the buds in all their glory. Some people dry their buds on the line first, while others get to trimming right away.

If you start with trimming, then you will find that the plants dry faster once they get up on the line. Right after you cut the plants down and bring them into your harvest area, you can identify the leaves that need to be trimmed because they will be filled with water. These leaves are firm and easy to cut. You should note that the trichomes are less likely to be knocked off at this point when you are handling the buds. This is because they have become more malleable. However, a trim tray can help you to catch what does fall off, and these can be used or sold later. If you trim your plants right away, the drying process will be shorter. This quick-drying method is one to consider if you need to have the plants dried quickly for some reason or if the location is too humid. If you are a low humidity environment, however, then it is a better idea to dry first and then trim because quick-dried buds in a low humidity may become harsh when smoked.

Drying the plants is very easy, but time consuming. Hang the plants up. You will find that after three to ten days, the stems have begun to warp and break slightly. When you see this, you will know that it is time to trim the buds. Dried leaves and branches are easy to trim off at this point because they pretty much just snap right off. However, the trichomes are also more likely to fall off. You can pretty much expect to lose a lot of trichomes at this point, so again it is a good idea to have a trim tray underneath.

Regardless of whether or not you dried or trimmed first, the trimming process remains the same. Use one hand to pick up the bud and maneuver it while using the other hand to cut away any leaves and branches. You may choose to do a tight trim really close to the buds, or you may choose to leave some filler there. If you are selling the buds to be smoked or consumed, then customers are going to expect a tight trim. Marijuana is sold by weight, and customers will not want to be paying for stems, which have no effect. So remove any large leaves sticking from the buds, along with any brown or yellow leaves. If a leaf doesn't have any trichomes on it, then you can get rid of it altogether.

There are a lot of tools that can help make your trim as easy as possible. Curved trimmers are fantastic for getting in and around the circular shape of a lot of the buds. There are also bowl trimmers, which have turning cranks that allow you to toss the bud inside and turn the crank to handle the trimming.

Take your time when you first start trimming. Do not just go in waving your clippers every which way. You will end up hacking up your bud and lose money in the process. It is better to start slow and get a feel for what is and isn't worth keeping. Razors are also a useful item to have around so that you can clean off your scissors, since they'll end up getting covered in sticky residue that can slow you down or clog up your scissors.

A Step-By-Step Guide to Harvesting Your Plants: Curing

Curing is a prolonged period of drying that can greatly improve your product. Plants create THCA and cannabinoids through a process called biosynthesis. Biosynthesis is a process in which some compounds are converted into new blends. For example, THCA turns into THC. You

might think this process stops when you cut the plant down, but in reality, the biosynthesis process continues after harvest. By putting the freshly harvested plants into 60°F to 70°F temperatures with a humidity level of about 50%, you will provide an environment in which THCA will continue to convert into THC. Do note that if you choose to go with the quick-drying method described above, then the biosynthesis process stops earlier than if you do not.

We talked about how terpenes provide the smell and flavor in Chapter Two. These terpenes are highly volatile and will quickly degrade or even evaporate entirely at temperatures of 70°F or higher. The slow-curing method helps keep these terpenes in place so that the marijuana continues to have its aromatic elements. The conditions for curing are also ideal for enzymes and aerobic bacteria to thrive, which breaks down minerals and sugars that arise from the decomposition of chlorophyll throughout the drying process. These elements can also cause the harsh and unpleasant taste and burning that quickly-dried marijuana often has. Curing also allows the cannabinoids to stay in place (without any worry about mold!) for a couple years before any loss of potency.

So you harvested your plants, cut the branches into 12 to 16 inch pieces, removed useless leaves, and then hung them on your wire. Keep them hung up in a dark room with the previously mentioned environmental conditions alongside a fan to keep the air gently circulating. You may also want to use a dehumidifier or air conditioner to ensure that the environment stays within these ranges. This first drying will need anywhere from five to fifteen days, after which the dried buds can be cured.

To cure your buds, first trim and dry the plants as described before. Here's where things get different. Take the trimmed buds and stick them into an airtight container. It is common to see mason or canning jars used for this, but anything airtight will work. You want the buds to be loosely packed into the containers rather than filling it all up. This is done so the buds aren't crushed in the process. Pack in some of the leaves you trimmed off. Seal the containers once they are full, and store them in a dark spot. You will want it to be cool and dry. After a day, you will notice that the buds have regained some moisture and are less "crunchy." This is because the remaining moisture in the flowers have rehydrated the bud. If this doesn't happen, then your plants were overdried earlier.

After canning your buds, you are going to want to open the containers a few times each day for the next seven days. This is done to allow them to breath. Basically, moisture will leave and oxygen will rush in. If you smell ammonia when you first open a container, then you underdried the plants. The smell arises from the anaerobic bacteria eating the organic matter, and you will likely find mold or rot on your buds as well. If you find that you underdried your plants and this situation occurs, you will have to throw out the buds. If not, after that first week, you can switch to opening the jar every other day or so instead of multiple times a day. In two or three weeks, you will find that the marijuana has cured enough to have a high quality. However, if you wait four or five weeks, then you will find it to be even better off. Some strains may even be best when cured half a year or more!

This simple step is often ignored, primarily because it slows down how quickly growers can sell or consume their product, but it is how you get the absolute best buds possible from your harvest.

Chapter Summary

- Harvesting lets you calculate how much your yield is worth and start the process of either enjoying or selling your product. For many growers, harvesting is the only part of the process they truly care about.

- When you harvest your crop will determine how potent and what kind of effect the bud will have.

- The amount of THC in the bud is determined by the color of the trichomes, which are little specks of white crystal on the bud that holds the THC.

- Translucent trichomes aren't mature yet and do not have enough THC to interest buyers.

- Cloudy trichomes are ready to be harvested and will provide a heady high.

- Amber trichomes are overripe but will provide a body-focused high to the user.

- If you want for cloudy trichomes to start turning amber then you can get a high that is both body and mind and can be worth a lot of money.

- Trichomes should be checked under a microscope, a pocket scope makes for a great tool in this so you do not have to take the bud off the plant.

- Prepare the harvest area. You want it to be 70F degrees or less.

- Make sure you have gloves, a cutting tool, ropes to hang the plants and a trim tray.

- Cut away any old leaves left on the plants, focusing on those that are the lowest as they just steal away energy from the bud.

- Stop spraying your plants two weeks before harvesting.

- Flushing your plants means to spray them with water to clean off any chemicals that may be on them. Start flushing a week before you plan to harvest. For plants in soil you spray them as often as your treatments. For hydroponics you switch out the reservoir with clean water every three days.

- Avoid strong lights when harvesting.

- You can cut the whole plant down when you harvest or you can harvest the ripest buds first and let the others wait a week before harvesting. Cutting the whole plant is faster but with a smaller yield, the other takes much more time.

- Cut down the plants and hang them up on your lines to dry.

- Trim away the leaves and stem, cleaning up the buds. Anything without trichomes can be cut away.

- Hang the plants up to dry. Drying before trimming makes for easier trim. Trimming before drying makes for quicker dry.

- Curing your buds by drying them properly and then storing them in containers to maintain their freshness. Curing takes more time to accomplish but makes your bud much richer.

In the next chapter, you will learn about the many annoying problems you may find yourself having to tackle to keep your plants healthy and pest-free. Pests, fungi, and mold are all dangers that growers must keep in mind and be vigilant against when raising their crops.

CHAPTER EIGHT

MAINTENANCE AND PREVENTATIVE CONTROL

From planting the seeds to harvesting the plants, the journey that you have taken in this book so far has explored the best-case scenario of growing cannabis. Unfortunately, there are many steps to maintaining the health and wellbeing of your plants. Pests, fungi, and mold are all threats that can easily destroy your crop if you aren't careful.

The best approach to maintaining and defending your crops from harm is to take a preventative approach. Rather than wait for your plants to be afflicted by one of the following annoyances, take preventative steps to keep them from taking up home on your plants in the first place. This will save you the time, money and frustration that comes from battling a major infestation.

Pests

There are a shocking number of pests that like to feed off of the cannabis plant. These range from large to small, some easy to spot and others masters of camouflage. Hopefully you will never have to encounter any of the following bugs.

Aphids: These are soft-bodied green, white, yellow, red, black, or brown bugs found all over the world. Baby aphids look like little white lines moving around on the leaves of your plants. They normally grow into round green or yellow creatures, but aphids infesting marijuana plants have been known to grow into black ones as well. Aphids can grow wings, and this has lead some growers into thinking they had an infestation of flies rather than aphids.

To eat, aphids stab cannabis leaves with their mouths and suck out the juices found on the inside. They like to make their homes on the underside of the leaves, so one way of spotting them is to run a paper towel or piece of toilet paper under your leaves and see if it comes away with streaks of blood. You can tell that one of your plants is dealing with a serious aphid problem when its leaves change into a yellow color and then begins to wilt.

An aphid infestation begins when winged aphids arrive, looking for a place to lay eggs. These pregnant scouts might not eat from the plant and so seem harmless, but they are usually the warning sign of a coming infestation. Preventing aphids isn't easy, especially if you grow your plants outdoors. Aphids take just a little over a week to mature, so they can quickly take up home. To get rid of them, you have to regularly look for signs of infestation and remove the bugs as soon as you see them. You can additionally use an insecticidal soap or even neem oil to get rid of them. Ladybugs eat aphids, and introducing them to your garden is a good preemptive measure to take. In addition, make sure to remove ants, as some species are known to farm aphids and deliberately increase their numbers.

Barnacle Scales: These weird-looking bugs like to take up a spot on the stems of the cannabis plant and then stop moving. They bite down into the stem and then start to weaken the plant by eating the juices inside. In doing so, the plant releases a substance called honeydew (which Aphids also love) that can attract ants and lead to mold patches on your plants. This means that the barnacle scales weaken the plant, increasing the risk of aphid infestation and mold, all at the same time.

These insects do not move around a lot, so they can be a little hard to spot. To make it more challenging for the grower, barnacle scales can also blend in and look like part of the plant itself. If you spot anything on your plant that looks a little odd, give it a splash with some water and see what happens. You want to use a water sprayer to knock the scales away. Follow this up by washing the plant with insecticidal soap and neem oil like you would for aphids. You may also want to introduce ladybugs, lady beetles, or lacewings into the garden to eat the scales.

Note that the scales reproduce very quickly, so they can easily double or triple in number in no time at all. This makes them a significant threat, because they take up hold and open your plants up for all sorts of other issues. Even worse, barnacle scales do not cause any noticeable symptoms themselves. They make it easier for other creatures while hiding out and moving as little as possible. This makes them an annoying but dangerous pest.

Broad Mites: Broad mites are microscopic, and sometimes hard to spot even when they are held under a microscope. The big problem with broad mites is that because you cannot easily see them, the symptoms of

their infestation are often thought to be stress, heat, overwatering, an imbalanced pH, or problems with the root. Symptoms of broad mites include new growths sticking off your plants that are twisted and droop weirdly, leaves that look wet or covered in blisters, or leaves that have the edges all turned up and wrong. Broad mites can also kill your buds when your plants begin to flower. These effects are uniform across the plant, but will always be worse in the spots where the infestation has been centralized.

The main way that people realize they have an infestation of broad mites is when they notice the damage these pests have caused. Since you cannot see them with your eyes, observing the damage is really the only way to clue into them. Their symptoms may look a lot like the effects of the tobacco mosaic virus.

Broad mites can be tricky to get rid of since many of the miticides out there aren't very good at killing them. Despite their size, broad mites are very tough. You have to cut away any part of the plant that has been infected. Any treatment you use to kill them should be used multiple times a day, and always spray the plants before you turn off the lights. You should continue any treatment you use for at least five weeks, even if you see signs of improvement in the plant before that. Broad mites can seem like they disappeared, only to show up again in a hurry because there was still one or two of them left. If you want to go the organic route and use a predator to lower the population, then you can add neoseiulus mites which prey on broad mites, into your garden.

Caterpillars and Inchworms: Caterpillars and pests like them leave some of the easiest to spot signs of their presence. If you start spotting large pieces of your leaves missing, then your first thought should be caterpillars. Clumps of brown stuff that looks like dirt on the leaves, or caterpillar poop, will confirm the infestation. Be aware that while caterpillars and inchworms look very similar, most worm-like pests on your plants are going to fall into the same categories as far as damage and treatment are concerned.

Getting rid of caterpillars is best done by using a biological insecticide such as Caterpillar BT Spray. This spray uses a bacteria called *Bacillus thuringiensis* to kill the caterpillar larva. It also makes it so that caterpillars cannot eat the cannabis leaves. The best part of using BT Spray is that though it focuses on caterpillar elimination, it also is known

for killing fungus, gnats, and moths. However, BT Spray will not harm many of the beneficial insects you add to your garden. Start using BT spray as soon as you spot evidence of caterpillars on your plants, be it from damage, poop, or catching them in the act. Use it weekly.

It's always best to prevent pests from taking hold when you have the chance. If you want to avoid caterpillars then you need to start thinking of butterflies and moths as the enemy of your plants. While a butterfly on your plant might make for a beautiful picture, butterflies actually lay eggs on the plants. So while you are reaching for the camera, they may be dropping off an infestation of caterpillars that will eat your plants. Shoo those critters off whenever you see them.

Crickets: One of the loudest of the pests, you will pretty much be able to tell if you have a cricket infestation from the noise alone. You will also be able to clearly notice the damage they leave, since they chew little holes through the leaves. You can even find mole crickets attacking your plants down by the root. One or two crickets can very quickly become a major infestation as they breed very fast. However, crickets do not go after cannabis plants as much as grasshoppers or scales.

If you think that you have a cricket infestation, then you should check the plants at night, since crickets are nocturnal creatures. They are attracted to lights at night, and so turning them off to match the sun's natural schedule can help you avoid them. They are also attracted to the messiness of unclean leaf litter and biological waste; even the garbage can draw them in. You can use cricket traps to capture them, but if you have a full blown infestation, these traps will only help to reduce their numbers, not solve the problem as a whole. Insecticidal soap and neem oil are again very useful. You may also consider using a floating row cover to prevent crickets from having access to the plants. This netting goes over your plants, but still allows water and sunlight to get through to the plant. Floating row covers are useful in preventing pests, but can be an annoyance to use.

Fungus Gnats: Gnats have the appearance of small flies. They enjoy spending time in the soil around your plants, as their young like to grow in wet soil before moving up to the surface to snack on your plants. Gnats show up often on plants that are overwatered, so the first step you should take in prevention is to ensure that you are watering your plants in a healthy manner. When watering your plants, let the top inch

of the soil dry out before you water them again. Doing so will leave the gnats without the moist soil they need to raise their young, in which case they will move on to other plants. This practice also helps in getting rid of an infestation but it can take weeks to work.

First check for fungus gnats by using your eyes to spot the bugs themselves. The gnats will be visible before any damage is, so they are their own warning sign. As soon as you spot the gnats, examine your watering habits and see if you should be slowing down. Adult gnats do not do much damage to your plants as larvae do, but they can still spread diseases, which could then absolutely devastate a crop. However, the real danger is in the larvae they leave in your soil. Though these creatures are tiny, they feed primarily on the roots of the cannabis plant, which will weaken the plants and can cause serious damage if not contained.

If you are having problems with fungus gnat larvae, then you will notice what is called a "damping off" of the seeds. Seedlings may appear to just be weak, without any real sign as to why they are just dropping dead. Adult plants slowly start to wilt and turn yellow with nasty spots all over them. The leaves will appear as if they have a nutrient deficiency or as if the pH level was causing a problem. The growth of your plants will slow down and maybe even stop, and the yield of your plants will be lower than normal.

Taking care of the watering issue is the first step in dealing with gnats and their sticky larvae, but you may want to use sticky traps down by your roots. Yellow-colored traps, in particular, are highly effective as gnats are drawn to the color. The glue on the trap will keep gnats stuck and help kill off the population. It also is a great way of seeing just how bad the infestation is, since the more gnats in the trap, the bigger the problem. A fan blowing over your plants and soil can be beneficial as well, since it keeps the soil dry and makes it more difficult to fly around. Larvae in the soil can be killed with neem oil, and sprinkling diatomaceous earth over the top of the soil is a great approach to prevent further issues.

Grasshoppers and Locusts: These annoying critters love the taste of cannabis leaves. They will chew right through the leaves and then start living on the branches of your plant with an attitude like they just started paying you rent. They'll eat the stems as well, which can kill entire branches of your plants. This makes them an absolutely deadly pest, more than capable of killing off a crop if left unchecked. They need to be dealt with quickly, or else they can leave you miserable.

Check for grasshopper infestations often. In fact, you should really be checking your plants for signs of any infestation on a daily basis, just to stay safe and vigilant. Spinosad products, which kill grasshoppers on touch, can be sprayed over the plants to reduce their numbers. Spinosad products are harmless to humans, pets, and plants, as they are made from the fermented actinomycete *Saccharopolyspora spinosa*. *S. spinosa* is a soil bacterium that causes the grasshoppers to die after ingestion by destroying their nervous systems. Insecticidal soaps and neem oil are effective here as well, as with many of the various pests you will encounter. Floating row covers help prevent grasshoppers, too, as the insects are too big to squeeze under the covers.

Leafhoppers: These little guys chew on the leaves of the cannabis plant, and in turn, leave the leaves covered in brown and yellow spots that resemble acne. Leafhoppers are also known for spreading disease, and this makes them particularly dangerous to your crops. There are many, many different kinds of leafhoppers. In fact, there are more than 20,000 identifiable types. This means that leafhoppers come in all shapes and sizes, though the most common type is the *Graphocephala coccinea.* Otherwise known as the candy striped leafhopper, these little colorful bugs would be beautiful to look at if they were not so dangerous. Leafhoppers like to eat your plants when the atmosphere is dry, since they are after the moisture in the cannabis leaves.

The spots that leafhoppers leave behind will be concentrated into small clusters. These insects suck sap from the leaves, which changes the leaf color in that location. They do not chew holes through leaves themselves, but rather just leave them looking gross and damaged. With so many different types of leafhoppers, they can look very different from one another. But all leafhoppers share a few attributes. Leafhoppers are insects, so they have six legs; they also all have wings and move by jumping and sliding. They also all die the same way, which is a great benefit to us growers.

Check your plants for signs of leafhoppers. If the atmosphere is dry then you should be checking for the little pests more often, as this is when they are the thirstiest. They like to hide on the underside of leaves, since they do not like to be caught snacking. Use spinosad products and follow with insecticidal soap and neem oil. Leafhoppers are food to lady beetles (such as ladybugs), lacewings, and parasitic wasps. These bugs will each eat quite a lot of leafhoppers, so releasing them can be a great way to reduce their numbers. They are also slowed down by floating row covers but these only work for prevention, not reduction of an already established infestation.

Leaf Miners: By far one of the weirdest pests that you can find yourself dealing with, the leaf miner actually lives inside of the cannabis leaf, rather than eating it or chewing on the outside like most other pests. Leaf miners open up the leaf to enter inside, and then they start to mine their way through it. This lets them eat the most delicious pieces of the leaves. You will be able to spot them inside your leaves as they bulge out. They also leave a trail of damage behind them that could look like

zigzagging white lines that run through the leaves. Leaf miners aren't so much a single species of pests as they are a catchall term for many different pests that use the leaves as a home and snack bar.

The first step in getting rid of a leaf miner problem is to cut away any of the leaves that have already been affected. If you have started to see the tell tale lines of damage they leave behind, then you should cut away these leaves immediately. This is highly unlikely to fix the problem outright, but it can reduce the number of leaf miners in the plant. You can crush the larva inside of the leaf itself if you are nervous about cutting them off. Spinosad is useful here, as is neem oil. A BT spray may help if the leaf miners are caterpillar larvae. Another option, after getting some floating row covers to reduce infection, is to release some *Diglyphus isaea* into your garden. These are parasitic wasps that will kill and eat leaf miners for you. These wasps like other pests too, and will eat them as well, but they'll eventually fly away after a few days, in which case you will have to purchase and release more.

Mealybugs: These bugs are covered in white hair that makes them look kind of like mold. Many growers see them and think they have a mold problem, and so they then treat their plants for the wrong issue. Mealybugs love warm temperatures and absolutely hate the cold. Worse, they do not just go after your plants while they're growing. Mealybugs can be found on your plants after harvest, making them one of the pests that hold on the longest. The patches they leave can look like mold, powder, or even the webbing of a spider. Just like scales and aphids, these bugs suck out honeydew from the cannabis plants, an act that leaves your plant susceptible to soot, mold, or ants (which makes the aphid problems worse). Mealybugs look like white versions of scales, except that they move a lot.

An infestation of mealybugs is going to be a pain to get rid of. Insecticidal soaps help in this regard, but as mealybugs are highly resilient critters, it is best to try more than one brand or method of elimination when tackling them. It will likely take constant treatment to get rid of mealybugs, but, as tempting as it might be, you should not use all of your chemicals at one. Rotate them one at a time instead. If you are using insecticidal soap on Monday, then you would use neem oil on Tuesday, for example. First, however, you should start by using a water spray to remove as many of the mealybugs as you can see. This is a step you can take every day, but combine it with a process like neem oil rather

than use the water spray to replace a day's step. When it comes to insecticidal soaps, it is best to find one that has a high concentration of fatty acid salts because these weaken the outer shell of the mealybug without damaging the plants or leaving behind a trace. Neem oil helps, but never get the stuff on your buds. Alcohol is actually a great way to kill them and you can either use a cotton swab to kill the bugs individually or you can mix rubbing alcohol with water to make a spray. Just keep it at nine to one water to alcohol mixture.

Of course, mealybugs are also super tasty to lady beetles, ladybugs, and lacewings, so these insects can be highly useful if you find you have a mealybug problem. Diatomaceous earth will work as well. This stuff is made from fossilized shells so it is very, very sharp when you look at it on the microscopic scale. This allows it to stab into and through the exoskeleton of pests, making them bleed out and lose the fluids inside of them, but it is too small to do any damage to larger animals like pets or children. In fact, while it would be gross, it could even be eaten without causing harm. Not that it is recommended, of course.

Slugs and Snails: These little guys love eating the leaves of your cannabis plants and even chewing on the buds. You may notice trails of slime on your plants that leads to or away from damage and holes that have scalloped edges. The bite marks from slugs and snails looks quite similar to the damage left by caterpillars. You will know it is slugs and snails instead of caterpillars if the edges of the damaged parts begin to look smooth. Like crickets, these pests are most active at night. They are more likely to appear in spring, and can come out earlier than many of the other pests. If they are attacking seedlings, then you shouldn't be surprised to see entire seedlings die overnight thanks to these little guys.

To deal with slugs and snails, you could allow frogs or beetles, their natural predators, into the garden. But even better than killing slugs and snails off after they've set up shop is to prevent the infestation by making it impossible for the slugs to reach your plants. You can do so by taking a plastic bottle, cutting it in half widthwise, and then using each half as a wall or makeshift greenhouse to cover a seedling. Another approach is to sprinkle lime, eggshells, diatomaceous earth, or sawdust in a circle around the plant. The slugs and snails will not be able to cross that barrier to get at your plant without killing themselves. Slugs seem to love

plates, orange rings, and beer. Fill a pot with some beer, and you can find slugs that dropped in and drowned themselves in the brew.

A really effective beer trap can be made by taking some flour and some stale beer and mixing them together in a container. Set this in your garden with the top a couple centimeters from the ground so that the slugs and snails can easily climb into the container. If you do not have beer, then you can use wine, juice, yeasty water, or even sugar water as the liquid part of the trap instead. The flour thickens the liquid and makes it almost impossible for the pests to escape. You can expect these traps to fill up often, so remember to check them on a daily basis.

Spider Mites: As suggested by their name, spider mites are part of the same overarching family that gives us spiders and ticks. They're one of the more common pests when it comes to cannabis crops. Unfortunately, they're also one that latches on hard and is hard to drive off. These pests have really sharp mouths which they use to bite into the cannabis plant and suck out its liquid. The leaves of your plants are left covered in tiny, bump-like yellow, orange, or white specks. Spider mites are more common when you are using a soil-based growing method, though they can still infect a hydroponic setup. One of the biggest problems with these pests is their small size, which allows them to really populate a crop before a grower notices them.

These pests reproduce quickly, seem to disappear only to come back a few days later, eat large volumes of plant matter, cover your plants in gross silk webbing that can further damage crops, and survive almost anything you throw at them. That's the worst part with them—spider mites quickly grow immune to your chemical solutions, so you need to switch up what you use to drive them off. Otherwise, they learn to ignore it. If you are growing indoors, then spider mites were most likely transmitted to your plants because you forgot to clean yourself off before you started working on your crop.

If you have an infestation of spider mites, then you will want to lower the heat and get some fans blowing on your plants. Spider mites like warmth, and they hate the wind, so make them regret making your plants their new home. You can spray water on them to knock them off as well. Next, use Azamax in small amounts with your watering to kill any mites that are in the soil. Spinosad should be used, along with Essentria IC3 insecticide, which is often used to kill bed bugs.

Insecticidal soaps with fatty acid salts will help. You will even want to get a specialized product like Dokto Doom Spider Mite Knockout Spray. Seriously, eradicating spider mites takes a lot of work because of their high immunity. Ladybugs and predatory mites may also help, along with diatomaceous earth. Remember to treat the entire growing area, all of your soil, and everything. They do not just hang out on the plants.

Only use one of the above chemical solutions at a time. Use one solution, wait two days, and then use another one. Keep this up until you think all of the spider mites are gone. Then do it again, repeating so that you use each of your chemical solutions once more. Spider mites are notorious for sticking around long after you thought they left, so it is better to play it safe than to be sorry.

Thrips: It really is annoying how so many of these pests are really tiny. Thrips are no different here. They are small and quick when adults but tiny and still when young. There are many different types of thrips. Some are gold with wings, others are dark, and some are wormy. No matter how they look, thrips all want to chew your leaves and suck out their insides to leave silver and bronze patches all over. The damage they leave looks similar to the damage made by spider mites, except a lot bigger and less uniform in shape. If left unchecked, then thrips will soon kill off the leaves they're eating.

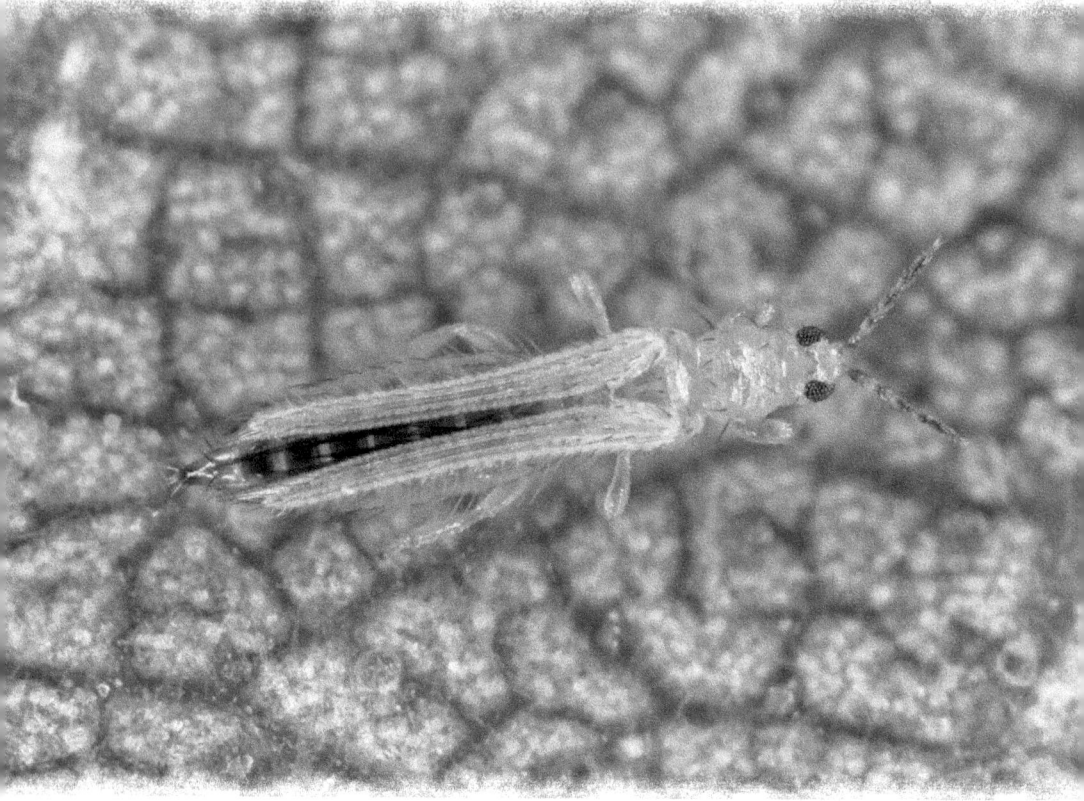

You're going to want to start treating the plants with insecticidal soap and make sure that you avoid any buds on the plant. Next comes the neem oil so that the pests find that your plants taste horrible, making them want to move on to tastier treats. Spinosad should come after that, so you can kill off a bunch of the thrips on contact. Another insecticide you can use for thrips is pyrethrins. However, pyrethrins are toxic to bees and some mammals, and so should only be used if you have no other options remaining. Hopefully, you will not need to worry about pyrethrins, as thrips are not nearly as hard to get rid of as spider mites are.

Whiteflies: The last of the pests, these ones look like white moths but act very similarly to spider mites. As with many other pests, they tend to collect on the bottom of leaves to suck their meal out. But though whiteflies eat on the bottom, it is the top of the leaf that starts to show white spots. You will likely notice the pests themselves, as their white color stands out against the green of the cannabis plant. Because of the way they flutter and move, you will most likely notice them before you notice the damage. Shake the plant and see if any whiteflies start to flutter around. This is the clearest sign in the world that you have an infestation on your hands. They lay eggs onto the leaves with a sticky residue that makes it almost impossible to wipe them off. This lets a population get out of hand quickly.

Start with your insecticidal soap and then move onto the neem oil. From there you can move into spinosad. Essentria IC3 is useful in killing off whiteflies when used daily. You need to treat your plants often as whiteflies can flutter away from the plant during treatment and land on another part that hasn't yet been attacked. The shifting nature of an infestation of whiteflies can make for a very annoying battle, so always be careful to treat all of your plants and not just the parts with the heaviest concentrations.

Fungi and Molds

Fungi show up when the humidity around your plants is too high or when your plants have been weakened by pests. Unlike pests, fungi do not move around and make themselves noticed. They instead show up suddenly and silently. Fungi might take some time getting settled but can spread quickly.

Prevention is better than treatment, so ensure that you are always following proper hygiene rules when dealing with your plants. Always disinfect the area between crops and clean your hands and clothes before entering the grow room. Always remove dead plant matter from around your crops as these grow, since the decomposition attracts fungi. You want to make sure the temperature stays in healthy ranges for the plants rather than the fungi. You can purchase organic products with fungicidal properties.

One common type of fungus is botrytis, or gray mold. This pathogenic fungus is easy to detect, but spreads quickly and can kill a plant in a couple of days. Look for the brown and gray, dry texture. Cut

off any infected part, taking a little bit of the healthy plant nearby to be safe.

Powdery mildew is a white, dust-like fungus that covers the leaves and buds of your plants. It's not very lethal, but should still be removed, so discard any infected areas and use fungicide on your plants.

Mildew will take hold on leaves, stems, or flowers. It is found on the underside of the plant and can be spotted because of the yellow stains it leaves on your leaves. If you notice yellow stains and white dust, then you have mildew. The first thing to do is to create better ventilation for the plants by cutting away the infected areas and opening up the plant to breathe better. Then, use your fungicide.

Fusarium is a fungus that is found not on the leaves but down in the soil. Fusarium feed on the plant until it dies by preventing sap flow inside the plant itself. When sap doesn't flow properly, the plant will start to rot at the stem, and you will notice the oldest of the leaves start to strain. This is bad news because there is no treatment for fusarium, so all there is to do is toss out the infected plants.

Chapter Summary

- Aphids are small bugs that like to hang out on the bottom of leaves and can be a pain to get rid of.

- Barnacle scales are pests that take up a spot on your plants and then do not move. While they do not do an overly lot of damage themselves, they open your plants up to wider threats.

- Broad mites are impossible to see with the naked eye and easy to misdiagnose. They're tough to kill so keep an eye out for their damage early.

- Caterpillars can absolute wreck havoc on cannabis. It is best to use a biological insecticide like Caterpillar BT spray to kill them.

- Crickets can be heard, making them easy to spot. They are attracted to messiness so keep your grow area clean.

- Fungus gnats love plants that are being overwatered. Make sure you water properly and you will probably never have to deal with them.

- Grasshoppers absolutely love eating cannabis and can kill off plants quickly.

- Leafhoppers come in all sorts of forms but they die the same. If the atmosphere is dry then you should be keeping an eye out for them.

- Leaf miners live in the leaves, requiring you to cut them out and throw them away.

- Mealybugs look like mold but are one of the hardest pests to get rid of.

- Slugs or snails can be beaten with beer traps.

- Spider mites are the worst pests because they quickly grow immune to insecticides.

- Thrips can also get out of control quickly but are much easier to deal with than spider mites.

- Whiteflies can be spotted because they fly away when disturbed. Make sure to treat all of the plants and not just the most infected parts, otherwise they'll just fly to new leaves.

- Fungi and molds are silent killers that you must watch out for and treat immediately.

In the next chapter, you will learn about the most common mistakes that new growers make when they first start cultivating cannabis. Avoiding these mistakes will save you lots of time and trouble with your first crops.

CHAPTER NINE

COMMON MISTAKES TO AVOID

We looked at some of the mistakes that new growers make when it comes to cloning plants. Now it's time to look at those mistakes that are common across almost all new growers. If you can avoid falling into these harmful behaviors, then you can start your first crop with a leg up to ensure you get the best marijuana possible.

Not Bothering to Look after the pH Level

Every plant will have an optimal pH in which it best grows, but many people do not even bother trying to understand what a pH level is when they first start out. There is a ton of vocabulary that you learn when you begin growing anything, a lot of which isn't overly important. However, it is crucial that you listen to and pay attention to your growing medium's pH.

What pH does it determine just how many and which kinds of nutrients your plants will be able to use. If you have a pH in the wrong ranges, then your plants will not be able to fully make use of the nutrients that you feed it. There is much literature about pH levels and its importance out there, but when it comes to growing marijuana, what you need to know is that cannabis grows best in a hydroponic medium when the pH is between 5.5 and 6.5 and best in a soil medium when the pH is between 6.0 and 7.0. Check your pH levels often to make sure everything is working properly.

Stop Overfeeding Nutrients

Nutrients are good for us and for our plants, so more nutrients would be more good, right? While the logic seems to check out, reality tells a different picture. When you overfeed your plants with too many nutrients, they will eventually suffer from nutrient burn. It might not kill your plants, but it will leave them feeling rough and unhealthy, making them more at risk from pests.

Create a feeding schedule for your plants so that they always get their nutrients when expected. Only use a quarter or half of the recommended dosage, as many companies recommend using way more

than needed. If the plant looks like it doesn't have enough nutrients but the pH is in the ranges above, then you will want to feed a little more nutrients on your next dosage.

Stop Overwatering Your Plants

Overwatering your plants is going to make them more susceptible to pests, start drooping down, and, if you are really over doing it, even die outright. Check your growing medium before watering your plants. Stick a finger in an inch and see if it is dry. If it is, then water your plants. But if it isn't, then do not water them yet. You shouldn't water them too often, only when they pass the dry check.

Control the Climate Properly

When it comes to growing cannabis, it is all about the temperature and humidity. They both need to be in the right ranges if you want your plants to really flourish. A poorly managed climate will hurt your plants more than any pest will.

Temperature needs to be just right. When it is low, your plants do not grow as quickly or as big. If the temperature is low enough then your plants will not grow at all, they'll just die instead. But if the temperature is too high, then the plants will suffer from heat stress, and you will see the leaves curl up because they are trying to get shade from the sun. If left to suffer heat stress for too long, then the plants will die. Keep the plants around 75°F.

In general, the humidity should be low. You may want a higher humidity on certain plants under specific circumstances, as with the clones we saw earlier, but for the most part, humidity is best kept low. An overly humid environment leaves your plants at risk for mold, fungi, and pests. Keep in mind that the ideal humidity should also change with the plant's life stage. A seedling might like humidity around 60% but as it grows that should be dropped down closer to 40%.

Chapter Summary

- The pH level controls how many nutrients your plants can make use of.

- Cannabis should have a pH between 5.5 and 6.5 in a hydroponic system. Plants grown in soil need a pH between 6.0 and 7.0.

- Too many nutrients will cause your plants to suffer nutrient burn and this will greatly weaken them, making threats from pests even more dangerous.

- Too much water can kill plants and open them up to infection and infestation.

- Cannabis needs to be kept around 75°F degrees.

- Humidity should be around the 40% range for full-grown plants and 60% for seedlings.

FINAL WORDS

As cannabis is increasingly legalized around the world, the industry will only continue to grow. Now is the time to get onboard, to make preparations for this shift, and get in on the ground floor. With the book in your hands, you have a resource that you can return to anytime to remember how to keep your marijuana crops healthy and get the most out of them.

Cultivating cannabis is one of those fields where you can really dive in and do as little or as much as you like. The bigger your operation, the more you will make back upon harvest. But many growers only want a plant or two of their own and that's perfectly fine too. Whatever your goals are with your crops, remember to take care of them and check in with them everyday. They might just be plants but you should consider them your babies, put in the work to keep them healthy, and see them grow up into beautiful buds.

Experiment with breeding and cloning your plants to get the most potent buds you can. Maybe you will invent the next strain to win awards. You never know if you do not try.

CANNABIS
COOKBOOK

Delicious Edible Medical Marijuana Recipes for Beginners

By Tom Gordon

CHAPTER ONE

BENEFITS AND SIDE EFFECTS OF CANNABIS

Before we begin, let us first understand everything you need to about cannabis, starting with its benefits.

Benefits of Cannabis

Stimulates Appetite

Cancer patients, who undergo chemotherapy generally have a poor appetite and are not able to eat well. Cannabis helps them restore their appetite.

Reduces Pain

Cannabis can greatly help ease pain in patients who suffer from cancer, muscle spasms, chronic pain, terminally ill patients, etc. It also helps fight certain cancers like lung cancer and brain cancer.

Glaucoma

Helps treat glaucoma.

Parkinson's Disease

Helps reduce tremors.

Insomnia

It helps induce relaxation and sleep, especially in terminally ill patients.

Reduces anxiety and depression.

Alzheimer's Disease

Slows down the progression of Alzheimer's disease and can improve motor skills.

Bone Problems/Joint Pain

Helps with problems like arthritis pain and broken bones.

Diabetes

It can help prevent diabetes. It also helps control diabetes if you are a diabetic.

Nausea and Vomiting

Helps control nausea and vomiting.

Seizure Disorders Like Epilepsy

Helps to reduce seizures.

Autism, ADHD, ADD

Has a calming effect on patients with autism, ADHD, and ADD, especially when they are violent, acting out, or aggressive. Helps them to concentrate and focus better.

Side Effects

Long-term usage of cannabis can hinder the development of the brain. Memory problems and learning disorders are common in growing children. There is a tendency for schizophrenia in adolescents. Sometimes consuming cannabis can have an effect on medications you are taking. It is especially common for those adults who consume blood thinners.

There is always a chance of overdosing. When you consume edible cannabis, it takes longer to start showing the effect on your body (takes a couple of hours) than when you smoke it (it takes only a few minutes). So people tend to overdose when they do not find immediate relief. Overdosing can cause nausea, psychosis, panic attacks, hallucinations, etc.

Long-term ingestion of cannabis can cause irregular heartbeat, heart attacks, increased heart rate, low blood pressure, high blood pressure, etc. Your reflexes can become slower. You may feel less energetic and tired all the time.

A couple of weeks of cannabis use can make you dependent on it, which might further lead to addiction. Suddenly stopping the intake of Cannabis can cause severe withdrawal symptoms.

Consuming cannabis in bipolar disorder patients can worsen their disorder.

Immunity can be weakened.

Mental health may get affected. If you already have mental health problems, it can worsen. You may not be able to think or judge properly. You may suffer from memory loss, schizophrenia, hallucinations, or paranoia.

Eyes can become red or dry. You may have blurred vision. You may feel drowsy, fatigued, or dizzy.

You may have a sore throat or cough. Your taste buds may be affected.

Diarrhea or constipation is common. Overdosing can cause nausea and vomiting.

Some patients may have a skin rash.

Using cannabis with other medicines may have side effects.

CHAPTER TWO

CANNABIS HERBS & DISPENSARY

The different types of cannabis are:

1. Cannabis Sativa

2. Cannabis Indica

3. Cannabis Ruderalis

There are also male plants and female plants. There isn't much difference in the species of cannabis. The species are more or less the same. It is more about where it is grown because of climatic conditions; the plants have different height.

Cannabis Sativa

They are tall plants (about 10 - 15 feet in height) with lengthy, thin leaves and soft seeds. This variety goes well with cooking. It can be smoked as well. One generally feels happy and energetic after consuming it. It is high in THC. This variety of cannabis helps in overcoming depression and mood disorders. It is also good for those with ADHD. It is cultivated in equatorial regions like India, Mexico, Thailand, etc. The flowers take time to grow.

Cannabis Indica

These plants do not grow as tall as Cannabis Sativa. They grow in bushes, and no is more than 6 feet in height. They have short, rounded leaves and soft seeds. These plants can be grown indoors. The flowers do not take as much time to grow as Cannabis Sativa. These are cultivated in Asian countries that have short winters like Nepal, Afghanistan, and Lebanon and also in North African countries like Morocco. The buds and flowers grow in clusters and are sticky compared to Cannabis Sativa. Because of its healing properties, it is

generally recommended. It is of great help in pain relief. It has a calming effect and great for curbing anxiety and putting you to sleep.

Cannabis Ruderalis

These plants are even shorter than Cannabis Indica. The height of the plants is about 1 ½ - 2 feet. The flowering happens earlier (within 3 – 4 weeks of sowing the seeds) compared to Cannabis Indica. They grow naturally in Russia and are considered weeds. They are also cultivated in Central Europe, Eastern Europe, and Central Asia. This variety of cannabis is great for calming and relaxing.

In spite of having these differences, they all can be grown together. Hybrid varieties of cannabis are also being cultivated nowadays. In fact, hybrid cannabis is more commonly and popularly grown as compared to their individual 3 varieties.

Marijuana Dispensaries

When you need to buy cannabis, buy it from a state where it is legal. Make sure it is from a licensed dispensary. Buy the kind that suits your personal needs. If you do not buy the right form, it can affect your health. You need a doctor's prescription for consuming cannabis, so don't forget to carry any other documents that are required in the state that has legalized it. Registration at the dispensary is necessary.

While Buying Cannabis

Keep the following aspects in mind before buying cannabis.

- Buy only good quality cannabis. It should look good and have a fresh smell, but not all dispensaries allow you to touch or smell it. Do not buy anything that looks old or has an awful smell. Good quality cannabis is expensive.

- If possible, visit a dispensary to make your purchase rather than ordering home delivery.

- Before going to a dispensary or finalizing your dispensary, check the reviews on the Internet. If you find many negative reviews, look for another dispensary.

- When you visit the dispensary for the first time, find out where they procure the cannabis if they do not grow it themselves. Find out the types of cannabis they have.

- You need to be at least 18 years of age to visit the dispensary.

- Since you are using it for medical purposes, you need not pay taxes.

- Qualified budtenders at the dispensaries can help you choose the right cannabis.

Choosing the Right Dosage

Consuming the same quantity of cannabis can have a different effect on different people. The effect varies from person to person, whether men or women. So you cannot have a generalized dose. The same goes for the

withdrawal symptoms as well. It can vary from person to person. Dosage also depends on how frequently and for what purpose you use it.

Your medicinal practitioner will not generally prescribe the exact dosage. He will generally recommend that you start with low doses, taking into consideration your age, medical conditions, the other medications you are on, etc. The medicinal practitioner will also consider if cannabis will have adverse effects on your current medicines. Ingesting very small doses is called micro-dosing, which is may be equivalent to 1/20 of a normal dose. You need to wait for 4-5 hours before taking the next dose. To start, take no more than 3 doses in 24 hours. Make sure to drink lots of fluids. In a few days, you will know how it is affecting your body and whether it is helping your problem and how well you are able to tolerate it. You can now decide if you want to lower the dose or increase the dose. This is called dose titration.

In most patients, low doses work well, but higher doses can have an adverse effect. Measure the cannabis tincture (when mentioned in any of the recipes) using a pharmacy dropper.

CHAPTER THREE

BASIC CANNABIS RECIPES

Always consult your medical professional before consuming cannabis.

Always adjust the cannabis in the recipes to suit your personal needs.

How to Decarb Cannabis

Preparation time: 2 minutes

Cooking time: 35 – 60 minutes

Makes: As required

Ingredients:

- Cannabis (marijuana), as required

Directions:

1. To prepare oven and baking sheet: Place a sheet of parchment paper on a baking sheet. Set up your oven and adjust the temperature to 220° F. Let the oven preheat.

2. Grind the cannabis to a fine powder.

3. Spread powdered cannabis all over the baking sheet. Place the baking sheet in the oven and bake for 35-60 minutes. You can bake it for longer if desired.

4. Remove from the oven. This is what you call decarboxylated or decarbed marijuana/ cannabis. Follow this procedure when decarbed is mentioned in the recipe.

Cannabis Tincture with Grain Alcohol

Preparation time: 5 minutes

Cooking time: 0 minutes

Makes: about 1 ½ cups

Ingredients:

- ½ ounce cannabis, ground into smaller pieces, decarbed

- ½ quart grain alcohol like Everclear

Directions:

1. Sterilize a quart size glass jar or mason's jar.

2. Add cannabis into the jar.

3. Pour alcohol into the jar.

4. Tighten the lid and place it in a cool, dark area like in a cupboard for about 2 – 3 weeks. You need to shake the jar every 2 – 3 days.

5. Strain with a cheesecloth or a fine wire mesh strainer into a beaker or measuring cup with a spout.

6. Pour the tincture into dropper bottles. Place in the refrigerator or a cool place until use.

Cannabis Tincture with Coconut Oil

Preparation time: 5 minutes

Cooking time: 6 – 8 hours

Makes: About 1 ½ cups

Ingredients:

- 2 cups cannabis, ground into smaller pieces

- 2 cups coconut oil

Directions:

1. To set up double boiler: Take 2 pots of nearly the same (but not same) sizes such that the smaller one fits inside, the larger pot.; the smaller pot should not touch the bottom of the bigger pot. It should fit well inside it.

2. Pour enough water into the larger pot such that it is 1/3 full. The water should not touch the smaller pot. Place the bigger bowl over medium flame. Let the water come to a boil.

3. Add coconut oil into the smaller pot. Place the smaller pot inside the bigger pot.

4. Lower heat to low heat and let the water simmer. Let it simmer this way for 6 – 7 hours. This is another method for decarboxylation.

5. Remove the inner pot from the double boiler and let it cool.

6. Line the top of the storage container with a piece of cheesecloth. Pour oil into it and discard the cannabis.

7. Tighten the lid and store at room temperature. It can last for a month. For it to last longer, place it in the refrigerator. It can last for 3 – 5 months.

Canna-Oil (Cannabis Infused Cooking Oil)

Preparation time: 30 minutes

Cooking time: 3 hours

Makes: About 1 ½ quarts

Ingredients:

- 32 ounces cooking oil like coconut oil /olive oil/avocado oil/canola oil

- ½ ounce cannabis, decarbed

Directions:

1. Add oil into a saucepan and place it over low flame.

2. Stir in the cannabis when the oil is warm.

3. Cook for about 3 hours, stirring every 30 – 40 minutes. Make sure not to boil or simmer the oil. Turn off the heat for a few minutes if it starts boiling.

4. Place cheesecloth over a fine wire mesh strainer. Place the strainer over a large heatproof bowl.

5. Pour the oil into the strainer. Squeeze the cheesecloth to remove as much oil as possible. You can wear gloves to protect your hand from the hot oil.

6. Let the oil come to room temperature. Pour into an airtight container and use as required.

7. When cannabis is infused with olive oil, it is called canna-olive oil; when infused with coconut oil, it is called canna-coconut oil. Similarly, with the others. If a recipe calls for canna-oil, use any cannabis-infused oil that you prefer.

CBD Oil

Preparation time: 10 minutes

Cooking time: 3 hours

Makes: About 1 cup

Ingredients:

- 1 whole hemp plant (with high CBD and low THC level) chopped, ground

- 2 cups carrier oil of your choice like fractionated coconut oil, olive oil, etc.

Directions:

1. Place ground hemp plant in a canning jar. Pour carrier oil over it. Fasten the lid.

2. Pour enough water in a saucepan, about 3-4 inches in height from the bottom of the saucepan. Place a washcloth in the saucepan along with the jar.

3. Place the saucepan over medium flame.

4. When the water begins to boil, lower the heat and simmer for 3 hours. Add more water if it goes dry.

5. Shake the jar every 30 minutes. You can hold the jar with a pair of tongs to shake.

6. At the end of 3 hours, remove the saucepan from heat. Place a towel on top of the saucepan. It should be totally covered. Let it cool for 3 hours.

7. Repeat steps 3-6 once again, but this time let it cool overnight.

8. If you want stronger CBD oil, repeat steps 2-6 every day for the next 2 to 3 days.

9. When you have the oil of the preferred strength, pass the oil through cheesecloth into a dark glass bottle.

10. Store in a cool and dry area.

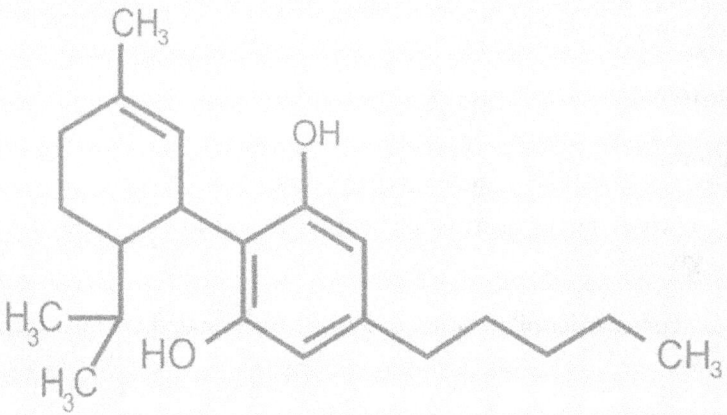

Cannabiol (CBD)

Canna-Butter

Preparation time: 5 minutes

Cooking time: 45 – 50 minutes

Makes: About ½ cup

Ingredients:

- ½ ounce cannabis buds, ground with a hand grinder into smaller pieces (do not powder it finely)

- 1 cup salted butter

Directions:

1. Place butter in a saucepan over low flame. When the butter begins to melt, add cannabis powder and stir often.

2. Let it simmer for about 50 minutes. Turn off the heat.

3. Strain into a glass container with a fitting lid. Press the residue with the back of a spoon. Discard the residue.

4. When the butter hardens, fasten the lid of the container.

5. Place the container in the refrigerator until use. If you want to make larger quantity of butter, cook for longer, approximately 60-80 minutes.

6. You can make canna- margarine or vegan butter similarly.

Canna-Honey

Preparation time: 10 minutes

Cooking time: 7 – 8 hours

Makes: About 1 ½ - 2 cups

Ingredients:

- 2 ¼ cups honey

- ½ ounce cannabis, decarbed

Directions:

1. Take a large piece of cheesecloth and place the cannabis in it. Bring the edges of the cloth together. Seal the marijuana well. It shouldn't come out of the cloth. Fasten with a string.

2. Pour honey in a crock-pot. Drop the cheesecloth bag in it.

3. Cover the pot. Cook on "Low" for 7 – 8 hours. Stir every hour.

4. Let the honey remain in the crock-pot overnight.

5. Heat the honey slightly, until just warm. Remove the cheesecloth bag and squeeze it with your hands and let the squeezed honey drop into the crockpot. Give it a good stir.

6. Store in airtight containers in a cool and dark place. It can last for 6 months.

Weed Sugar

Preparation time: 5 minutes

Cooking time: 2 hours and 30 minutes

Makes: 4 cups

Ingredients:

- 2 tablespoons citric acid (optional)

- 4 cups granulated sugar

- 2 cups cannabis tincture with grain alcohol

Directions:

1. Take a large, rimmed baking sheet and add sugar into it. Pour cannabis tincture over the sugar, stirring all the while. Add citric acid, stirring all the while.

2. Once well combined, spread the sugar evenly on the baking sheet.

3. Place the baking sheet in a preheated oven at 200°F. Point to be noted is that the oven door will remain open throughout the baking time. Also, all the windows and doors in the kitchen need to be open while baking.

4. Stir the mixture every 10-12 minutes for about 2 – 3 hours or until dry, and the sugar crystalizes once again.

5. Remove from the oven and cool completely.

6. Store in an airtight container.

Canna-Flour

Preparation time: 10 minutes

Cooking time: 0 minutes

Makes: 2 cups

Ingredients:

- 2 cups all-purpose flour

- ½ ounce cannabis, decarbed

Directions:

1. Finely grind the cannabis. Add flour and cannabis into a mixing bowl. Whisk until well incorporated. You can use a spatula or electric hand mixer or with a whisk attachment in your food processor.

2. Transfer the flour into an airtight container. Place it in a cool and dry place, like your cupboard, until use. It can last for 3 months.

Canna Milk (Cannabis Infused Milk)

Preparation time: 5 minutes

Cooking time: 45 minutes

Makes: https://theweedscene.com/cannabis-milk/ 4 – 5 cups

Ingredients:

- 1.8 ounces finely powdered cannabis

- 8 cups milk

Directions:

1. Place a pot with milk over low heat.

2. Add milk and simmer on low heat for about 35-40 minutes. Stir frequently.

3. Simmer until the milk reduces to nearly half its original quantity. The color of the milk is greenish with a tinge of yellow.

4. Strain through a fine wire mesh strainer placed over a bowl.

5. Serve hot or chilled.

Canna-Cream Cheese

Preparation time: 5 minutes

Cooking time: 30 – 40 minutes

Makes: About ¾ cuphttps://eatyourcannabis.com/canna-cream-cheese/

Ingredients:

- ½ ounce cannabis, finely ground

- ½ quart cultured buttermilk

- ½ gallon whole milk

- ¼ teaspoon salt

Directions:

1. Add milk, cannabis, and buttermilk into a saucepan. Place saucepan over medium flame. Stir occasionally until the temperature of the milk is around 170° F to 175° F on a candy thermometer.

2. Let the mixture now simmer between these temperatures for around 10 – 12 minutes. Turn off the heat when the milk curdles.

3. Place a couple of layers of cheesecloth inside a strainer. Place the strainer on a bowl.

4. Pass the curdled milk through the strainer. Retain the curds and use the milk that is remaining in some other recipe like a smoothie or discard it.

5. Cool the curds completely in the strainer.

6. Add cooled curds into a blender. Add salt and blend until smooth.

7. Transfer into an airtight container and chill until use. It can last for 3-4 days.

Cannabis Mayonnaise

Preparation time: 5 minutes

Cooking time: 0 minutes

Makes: 2 – 3 cups

Ingredients:

- 6 egg yolks

- 1 teaspoon Dijon mustard

- 2 teaspoons lemon juice

- 2 cups canna-oil cannabis-infused oil

- Salt to taste

- 2 teaspoons white vinegar

- Pepper to taste

Directions:

1. Add egg yolks, vinegar, lemon juice, salt, pepper, and Dijon mustard into a blender. Blend until smooth.

2. With the blender running blender on low speed, pour the canna-oil through the feeder tube in a thin stream until the mayonnaise is emulsified and thick.

3. If you find the mayonnaise very thick, add a teaspoon of water to dilute.

4. Transfer into an airtight container and refrigerate until use.

Cannabis Peanut Butter

Preparation time: 5 minutes

Cooking time: 0 minutes

Makes: 5 tablespoons

Ingredients:

- 3 teaspoons cannabis-infused extra-virgin olive oil

- 4 tablespoons peanut butter

Directions:

1. Add canna-oil and peanut butter (you can use either creamy or chunky peanut butter, your favorite brand) into a jar or bowl.

2. Mix well with a spoon until smooth and creamy.

3. You can spread it on bread slices or add it in smoothies or anything you like.

Cannabis-Infused Apple Cider Vinegar

Preparation time: 15 minutes

Cooking time: 0 minutes

Makes: 7 – 8 cups

Ingredients:

- 8 cups organic, unpasteurized apple cider vinegar

- 1 ½ ounces high-quality cannabis flowers

Directions:

1. Decarb the cannabis flowers, following the procedure mentioned in the first recipe.

2. Place cannabis in a large, glass gar. Pour vinegar into the jar. Fasten the lid and shake the jar constantly for about a minute.

3. Place the jar in a cool and dry area; say in a cupboard, for about 25 days. Shake the jar every 3 days.

4. Remove the lid of the jar.

5. Place cheesecloth over the rim of the jar. Fasten with a rubber band.

6. Now pour the vinegar into a pitcher. Discard the cannabis along with the cheesecloth.

7. Pour the vinegar back into the same jar.

8. Repeat steps 5 – 7 using new cheesecloth.

9. Cover the jar with the lid. Fasten the lid and refrigerate until use.

10. It can last for about 2 years in the refrigerator.

Marijuana Vinaigrette

Preparation time: minutes

Cooking time: 0 minutes

Makes: About 2/3 cup

Ingredients:

- ½ teaspoon minced garlic

- 1 teaspoon fresh basil and oregano mixture or ½ teaspoon dried basil and oregano

- ½ cup canna- oil

- Pepper to taste

- ½ tablespoon minced red onion or shallots

- 2 tablespoons balsamic vinegar or any other vinegar

- Salt to taste

Directions:

1. Place garlic, oregano or basil, onion, vinegar, salt, and pepper in a blender and blend until pureed.

2. With the blender running, pour cannabis-infused oil in a very thin stream. Blend until the vinaigrette is slightly thick and emulsified or until the consistency you desire is achieved. If your vinaigrette is not thickening, add more oil.

3. Add salt and pepper to taste.

4. Transfer into a bowl or jar. Cover and refrigerate until use. It can last for 4 – 5 days.

Italian Cannabis Dressing

Preparation time: 5 minutes

Cooking time: 0 minutes

Makes: ½ - 2/3 cup

Ingredients:

- ¼ cup canna-oil
- 2 tablespoons grated Romano cheese
- 3 tablespoons red wine vinegar
- ½ - 1 teaspoon sugar
- ½ teaspoon freshly ground pepper
- 1/8 teaspoon garlic powder
- ½ teaspoon dried basil
- ½ teaspoon dried oregano
- 1/8 teaspoon red pepper flakes
- Salt to taste

Directions:

1. Add canna-oil, cheese, vinegar, sugar, pepper, garlic powder, basil, oregano, red pepper flakes, and salt into a small glass jar.

2. Fasten the lid and shake the jar constantly for about 30 seconds.

3. Refrigerate until use. It can last for about 15 days.

4. Make sure to shake the dressing before drizzling on the salad.

Simple Salad Dressing

Preparation time: 5 minutes

Cooking time: 0 minutes

Makes: 1-½ cups

Ingredients:

- ½ cup canna-olive oil or canna- avocado oil

- 4 – 6 cloves garlic, peeled, minced

- Salt to taste

- 1 cup lemon juice

- Pepper to taste

Directions:

1. Add oil, garlic, salt, lemon juice, and pepper into a mason's jar.

2. Fasten the lid and shake the jar constantly for about 30 seconds.

3. Refrigerate until use. It can last for about 15 days.

4. Make sure to shake the dressing before drizzling on the salad. You can also top it on meat.

Sweet Salad Dressing

Preparation time: 5 minutes

Cooking time: 0 minutes

Makes: About ¾ cup

Ingredients:

- 1 tablespoon lemon juice

- Salt to taste

- 2/3 cup mashed berries of your choice

- 3 tablespoons canna-olive oil or canna-avocado oil

- Pepper to taste

Directions:

1. Add oil, berries, salt, lemon juice, and pepper into a mason's jar.

2. Fasten the lid and shake the jar constantly for about 30 seconds.

3. Refrigerate until use. It can last for about 2 days.

4. Make sure to shake the dressing before drizzling on the salad. You can also top it on meat.

Honey Mustard Salad Dressing

Preparation time: 5 minutes

Cooking time: 0 minutes

Makes: 1 – 1 1/3 cups

Ingredients:

- ½ cup apple cider vinegar

- 2/3 cup canna-olive oil or canna- avocado oil

- 3 – 4 teaspoons Dijon mustard

- Honey to suit your taste (you can also use canna-honey if you want more cannabis in your diet)

Directions:

1. Add vinegar, canna-oil, mustard, and honey into a bowl and whisk until well combined and smooth.

2. Cover the bowl. Refrigerate until use. It can last for about 5 – 6 days.

3. Make sure to stir the dressing before drizzling on the salad. You can also top it on meat.

Cannabis-Infused BBQ Sauce

Preparation time: 10 minutes

Cooking time: 15 minutes

Makes: 7 – 8 cups

Ingredients:

- 4 cups ketchup

- 1 cup apple cider vinegar

- 2/3 cup sugar

- 2/3 cup light brown sugar

- 1 tablespoon onion powder

- 2 tablespoons Worcestershire sauce

- ½ cup canna-oil

- 2 cups water

- 1 tablespoon freshly ground pepper

- 1 tablespoon ground mustard

Directions:

1. Add ketchup, vinegar, sugar, light brown sugar, onion powder, Worcestershire sauce, canna-oil, water, pepper, and mustard into a pot.

2. Place the pot over low flame. Stir often until sugars dissolve completely. Let it cook for about 18 – 20 minutes.

3. Turn off the heat. Transfer into an airtight container.

4. You can use it for BBQ chicken or meat as a marinade or use it as a dip or spread it over cooked meat as topping.

Hazelnut Spread

Preparation time: 10 minutes

Cooking time: 15 minutes

Makes: About a cup

Ingredients:

- 5.3 ounces cream

- 0.6 – 1 ounce canna-butter (it depends on the strength of the cannabis)

- 1 teaspoon organic hazelnut extract, at room temperature

- 5 ounces semi-sweet chocolate chips

- 2.8 ounces hazelnut butter, at room temperature

Directions:

1. Add cream and hazelnut butter into a saucepan. Place the saucepan over medium flame. Whisk using a hand whisk until well combined.

2. When the mixture begins to simmer, turn off the heat.

3. Place chocolate chips in a bowl. Pour hazelnut –cream mixture over the chocolate chips. Cover and let it rest for 3 – 4 minutes.

4. Once again, whisk the mixture using a hand whisk until smooth and chocolate chips dissolve completely.

5. Add hazelnut extract and canna-butter and whisk until smooth and well combined.

6. Continue whisking until butter melts.

7. Cover the spread with cling wrap, with the cling wrap touching the spread.

8. Chill until use.

9. To use: Remove the bowl from the refrigerator and keep it on your countertop for an hour before serving.

Cannabis Salsa

Preparation time: 15 minutes

Cooking time: 0 minutes

Makes: about 1 ½ cups

Ingredients:

- 1 cup chopped fresh tomatoes

- ¼ cup diced onion

- A handful fresh cilantro, chopped

- ¼ fresh jalapeño, diced

- Lime juice to taste

- Salt to taste

- ¼ cup diced bell pepper

- 2 tablespoons canna-oil

- Pepper to taste

Directions:

1. Add tomatoes, onion, cilantro, jalapeño, lime juice, bell pepper, canna-oil, and pepper into a bowl and toss well.

2. Cover and chill until use. It can last for 2 days.

Cannabis Caramel Sauce

Preparation time: 5 minutes

Cooking time: 30 – 40 minutes

Makes: 8 – 10 servings

Ingredients:

- 2 cans (14 ounces each) full-fat coconut milk

- 1 cup coconut sugar

- 4 teaspoons canna-coconut oil

- 1 teaspoon vanilla extract

- Salt to taste (optional)

Directions:

1. Add coconut sugar and coconut milk into a saucepan. Place the saucepan over medium flame.

2. Stir frequently until the mixture comes to a simmer. Stir frequently until thickened to the consistency you desire. Point to be noted is that the sauce will become thicker on cooling.

3. Turn off the heat and cool. If the consistency you desire is achieved on cooling, go to the next step else simmer for some more time.

4. Whisk in salt, canna-coconut oil, and vanilla. Keep whisking until oil is well combined.

5. Pour into a container and refrigerate until use.

Cannabis Glycerin Concentrate

Preparation time: 5 minutes

Cooking time: 24 – 26 hours

Makes: About ¾ cup

Ingredients:

- 1.8 ounces cannabis trim

- 1 cup food grade vegetable glycerin

Directions:

1. Decarb the cannabis as given in the first recipe. Roughly powder it and add into a bowl along with glycerin and stir.

2. Place the bowl in the slow cooker. Set the temperature to 180° F and time for 24 – 26 hours, stirring after every 3 hours. Use a wooden spoon to stir.

3. Take a large piece of cheesecloth and place it over a jar or storage container. Pour the glycerin concentrate into it. Bring together the edges and squeeze out as much glycerin as possible. Fasten the lid and chill until use.

4. You can use it instead of honey to sweeten smoothies etc.

CHAPTER FOUR

BREAKFAST RECIPES

Weed Crepes

Preparation time: 10 minutes

Cooking time: 10 minutes

Makes: 4 servings

Ingredients:

- 2/3 cup canna milk
- ½ cup all-purpose flour
- 1 tablespoon white sugar
- 2 eggs, lightly beaten
- 1 tablespoon butter, melted + extra to fry
- ¼ teaspoon salt

Directions:

1. Combine canna-milk, flour, sugar, eggs, butter, and salt in a bowl and whisk until you get a batter that is smooth and free from lumps.

2. Place a medium-sized pan over medium flame. Add a little butter or oil.

3.   Pour 3 tablespoons of the batter into the pan. Swirl the pan to spread the batter thinly. Cook until the underside is golden brown. Flip the crepe over and cook the other side until golden brown.

4.   Remove the crepe onto a plate.

5.   Repeat steps 2 – 4 and make the remaining crepes.

Cannabis Waffles

Preparation time: 10 minutes

Cooking time: 20 minutes

Makes: 2 – 3 servings

Ingredients:

- 2 tablespoons canna-butter

- 6 tablespoons sugar

- 1 large egg, separated

- ½ cup butter, melted

- 1 cup all-purpose flour

- 1 ¾ teaspoons baking powder

- ¾ cup whole milk

- ½ teaspoon vanilla extract

To serve:

- Fresh berries

- Syrup of your choice

- Whipped cream

- Any other toppings of your choice

Directions:

1. Add sugar, flour, and baking powder into a bowl and stir until well combined.

2. Beat the yolk in a bowl. Beat in the butter, canna-butter, milk, and vanilla.

3. Pour into the bowl of dry ingredients and stir until just incorporated, making sure not to overbeat.

4. Beat whites with an electric hand mixer until stiff peaks are formed.

5. Add the whites into the batter and fold gently.

6. Set up your waffle iron and preheat it following the manufacturer's instructions. Grease the iron with some butter or oil.

7. Pour some batter into the waffle iron (about ½ cup). Set the timer for 4 – 6 minutes or until the way you like it cooked. Remove the waffle and serve with any of the suggested serving options.

8. Repeat steps 6 – 7 and make the remaining waffles.

Weed French Toast

Preparation time: 10 minutes

Cooking time: 45 minutes

Makes: 8 servings, 2 toasts each

Ingredients:

- 2 French baguettes, cut each into 8 slices, crosswise, ¾ - 1 inch thick

- 6 tablespoons canna-butter

- 3 tablespoons butter, unsalted + extra for greasing

- 8 eggs

- 1/3 cup sugar

- 1 ½ cups milk

- 6 tablespoons maple syrup

- 2 teaspoons salt

- 2 teaspoons vanilla extract

- Powdered sugar to top

Directions:

1. Prepare a baking dish by greasing it with butter.

2. Add canna-butter and butter into a bowl and mix well. Spread this mixture on one side of each slice of bread.

3. Place the bread slices in the baking dish with the buttered sided facing up.

4. Whisk together eggs, sugar, milk, maple syrup, salt, and vanilla extract in a bowl until well combined.

5. Pour this mixture over the bread slices. Press it down with a spoon.

6. Cover and chill overnight.

7. Place the baking dish in an oven that has been preheated to 350° F and bake for about 45 minutes or until golden brown.

8. Remove the baking dish from oven. Cool for 5 minutes.

9. Sprinkle powdered sugar on top and serve.

Weed Pancakes

Preparation time: 10 minutes

Cooking time: 10 minutes

Makes: 8 servings

Ingredients:

- 2 tablespoons white sugar

- ½ teaspoon baking soda

- 1 ½ cup canna-milk

- 2 cups all-purpose flour

- 2 teaspoons baking powder

- 1 teaspoon salt

- 1 tablespoon canna butter, melted

- 1 tablespoon white vinegar

- Cooking spray

To serve:

- Honey or any other syrup to serve

- Berries to serve

- Whipped cream

- Any other toppings of your choice

Directions:

1. Mix together all the dry ingredients, i.e., flour, baking soda, baking powder, salt, and sugar, in a bowl.

2. Pour milk and vinegar into another bowl and let it sit for a few minutes, about 5 – 8 minutes.

3. Add eggs and butter and whisk well.

4. Pour butter - milk mixture into the bowl of dry ingredients and whisk well until it is smooth and free from lumps. Set aside for 10-15 minutes.

5. Place a nonstick skillet over medium heat. Spray with cooking spray.

6. Pour about 4 tablespoons of batter, you can use a ladle.

7. In a couple of minutes, bubbles will be visible on the top of the pancake. Cook until the underside is golden brown. Turn the pancake over and cook the other side as well.

8. Remove pancakes from the pan and keep warm.

9. Repeat steps 5 – 8 and make the remaining pancakes.

10. Serve with any of the suggested toppings.

Cannabis Granola Breakfast Bars

Preparation time: 10 minutes

Cooking time: 20 – 30 minutes

Makes: 20 – 25 servings

Ingredients:

- 1 cup canna-coconut oil (infused coconut oil), melted

- 2 cups chopped nuts

- 2 teaspoons baking soda

- 3 teaspoons ground cinnamon

- Flavoring of your choice (optional)

- 1 cup brown flaxseed meal

- 1 cup honey or maple syrup

- 1/8 teaspoon salt

- 6 cups oatmeal

- 1 cup berries or fruits of your choice, fresh or dried

Directions:

1. Prepare a large rimmed baking sheet by lining it with parchment paper. Also, prepare the oven by preheating it to 300° F.

2. Add oatmeal, flaxseed meal, nuts, and cinnamon into a bowl and mix well.

3. Add oil, salt, honey, and any flavoring, if using, into a bowl and whisk until well incorporated. Pour into the bowl of the oatmeal mixture.

4. Mix until well incorporated. Transfer the mixture onto the prepared baking sheet. Spread it evenly but do not press the mixture onto the baking sheet.

5. Place the baking sheet in the oven and bake for 20-30 minutes or until golden brown. Stir once halfway through baking.

6. Remove the baking sheet from the oven and mix in the berries. Spread the mixture again on the baking sheet, evenly.

7. Let it cool slightly. Make 20 – 25 equal portions of the mixture and shape into bars.

8. Transfer into an airtight container and refrigerate until use. It can last for 5 – 6 days.

Blueberry Muffins with Weed Streusel

Preparation time: 20 minutes

Cooking time: 30 minutes

Makes: 18 servings

Ingredients:

For muffins:

- 2 cups milk

- 8 tablespoons butter, at room temperature

- 2 eggs

- 4 2/3 cups flour

- A large pinch salt

- 2/3 cup sugar

- 5 teaspoons baking powder

- 2 teaspoons vanilla extract

- 3 cups frozen or fresh blueberries

For streusel:

- 2/3 cup all-purpose flour

- ½ cup chilled canna-butter

- 1 cup sugar

- 1 teaspoon ground cinnamon

Directions:

1. To prepare muffin pans and oven: Take 3 muffin pans of 6 counts each. Grease them with cooking spray. Place cupcake papers in the muffin pans.

2. Place rack in the center of the oven and preheat the oven to 350° F.

3. To make muffins: To mix the dry ingredients: Mix together flour, salt, and baking powder into a mixing bowl and stir well.

4. To mix wet ingredients: Add sugar and butter into a large mixing bowl and whisk with an electric hand mixer until creamy.

5. Add vanilla and mix well. Add eggs, one at a time, and whisk well each time.

6. Add dry ingredients into the bowl of wet ingredients. Also, pour the milk and mix until just combined, making sure not to overbeat.

7. Add blueberries and stir well.

8. Pour into the muffin molds, up to 2/3 of the molds.

9. To make streusel topping: Add flour, sugar, and cinnamon into a bowl and mix well.

10. Add canna butter and mix it well into the mixture of flour until crumbly in texture. Scatter this mixture over the batter in the muffin cups.

11. Place the muffin pans in the oven and bake the muffins for about 20 – 25 minutes or until light brown. Cook in batches if required.

12. Cool completely on a cooling rack and serve.

13. Blueberries can be replaced with strawberries or any other berries of your choice.

Weed Breakfast Casserole

Preparation time: 15 minutes

Cooking time: 45 minutes

Makes: 6 servings

Ingredients:

- ½ package (from a 16 ounces package) ground pork breakfast sausage

- ½ can (from a 10.75 ounces can) condensed cream of mushroom soup

- ½ can (from a 4.5 ounces can) sliced mushrooms

- ¼ cup shredded cheddar cheese

- 6 eggs

- ½ + 1/8 cup canna-milk

- Salt to taste

- ½ package (from a 32 ounces package) frozen potato rounds

- Pepper to taste

Directions:

1. Place a skillet over medium-high flame. Add sausage and cook until brown, stirring often.

2. Meanwhile, prepare a baking dish by greasing it with some cooking spray. Prepare the oven by preheating it to 350° F. Place rack in the center of the oven.

3. Add eggs, milk, and mushroom soup into a bowl and whisk until well incorporated.

4. Add mushrooms and sausage and stir well.

5. Place the potato rounds in the baking dish. Spoon the sausage mixture into the baking dish and place the dish in the oven. Bake until set and light brown on top. It should take around 35 – 40 minute. Anyway, keep a watch over it after 30 minutes of baking.

6. Scatter cheese on top and place the baking dish back in the oven. Bake for another 5 to 7 minutes or until cheese melts.

7. Serve hot.

Breakfast "Baked" Burritos

Preparation time: 10 minutes

Cooking time: 15 minutes

Makes: 2 servings

Ingredients:

- 3 ounces bacon

- 2 eggs

- 1.5 ounces cheddar cheese, shredded

- 1 tablespoon canna-butter

- 6-7 tablespoons refried beans

- 2 flour tortillas (10 inches each)

Directions:

1. Place a large deep skillet over medium-high heat. Add bacon and cook until brown.

2. Remove with a slotted spoon and place on a plate lined with paper towels. Chop the bacon into smaller pieces.

3. Warm the tortillas following the instructions on the package.

4. Place a nonstick skillet over medium heat. Add canna-butter.

5. When the butter melts, crack the eggs and fry the eggs according to your preference.

6. Spread refried beans over the warm tortillas. Place half the pieces of bacon and an egg on each tortilla.

7. Sprinkle cheese on top.

8. Wrap like a burrito and serve.

160

Veggie Frittata

Preparation time: 10 minutes

Cooking time: 20 minutes

Makes: 4 servings

Ingredients:

- 4 large eggs

- Sea salt to taste

- 1 tablespoon cannabis-infused olive oil or hash oil

- 1 small onions, thinly sliced

- 2 ounces feta cheese, crumbled

- 3 tablespoons milk

- Pepper to taste

- ½ medium red bell pepper, thinly sliced

- 1 cup baby spinach leaves

- Fresh basil leaves for garnishing (optional)

Directions:

1. Whisk the eggs well. Add milk, salt, and pepper. Whisk until well combined.

2. Add cannabis-infused olive oil to an ovenproof skillet. Place the skillet over medium flame.

3. When oil is heated, add onions and bell pepper and sauté until slightly tender.

4. Stir in the spinach and sauté until the spinach wilts.

5. Pour the egg mixture over the onion mixture. Cook for about a minute.

6. Sprinkle feta cheese on top of the egg layer, making sure not to stir.

7. When the sides are cooked and the middle undercooked, turn off the heat.

8. Shift the skillet into an oven that has been preheated to 350° F and bake for about 15 minutes. Set the oven to broil mode and bake for another 5 minutes or until or until very light golden brown on top.

9. Take out the skillet from oven. Cool for 5 minutes.

10. Garnish with fresh basil leaves. Cut into wedges and serve.

Kale and Chevre Goat Cheese Frittata

Preparation time: 15 minutes

Cooking time: 25 minutes

Makes: 4 servings

Ingredients:

- 4 large eggs
- Sea salt to taste
- 1 ½ tablespoons cannabis-infused olive oil or canna-butter
- ½ large shallot, thinly sliced
- 3 mushrooms, sliced
- 2 small cloves garlic, peeled, thinly sliced
- 3 tablespoons crumbled soft goat's cheese
- Pepper to taste
- 1 cup baby kale leaves
- 4 tablespoons almond milk
- ½ small red onion, chopped

Directions:

1. Add ½ tablespoon cannabis-infused olive oil to an ovenproof skillet. Place the skillet over medium flame.

2. When oil is heated, add shallot, onion, and garlic and sauté for a minute.

3. Stir in the mushrooms and cook until slightly tender.

4. Stir in the kale and sauté until the kale wilts. Remove the vegetables onto a plate. Clean the skillet with a paper napkin and place it back over medium flame.

5. Add remaining oil into the skillet and let it heat. Swirl the pan to spread the oil.

6. Whisk the eggs well. Add almond milk, salt, and pepper. Whisk until well combined.

7. Pour the egg mixture into the skillet. Scatter the onion mixture all over the egg. Cook for about a minute.

8. Sprinkle goat's cheese on top of the vegetables, making sure not to stir.

9. When the sides are cooked and the middle undercooked, turn off the heat.

10. Shift the skillet into an oven that has been preheated 350° F and bake for about 15 minutes. Set the oven to broil mode and bake for another 5 minutes or until or until very light golden brown on top.

11. Take out the skillet from oven. Cool for 5 minutes.

12. Cut into wedges and serve.

Weed Omelet

Preparation time: 5 minutes

Cooking time: 10 minutes

Makes: 2 servings

Ingredients:

- 8 large eggs

- ½ teaspoon pepper or to taste

- ¼ cup chopped red bell pepper

- ¼ cup chopped green onion

- ½ cup shredded cheese

- ½ teaspoon salt or to taste

- ¼ cup cooked diced ham or any other meat of your choice

- 2 tablespoons butter

- 1 cup canna-milk

Directions:

1. Add eggs into a bowl and whisk well. Add milk, salt, and pepper and whisk well.

2. Stir in the ham, bell pepper, and green onion.

3. Place a skillet over medium flame. Add 1 tablespoon of butter and let it melt. When butter melts, pour half the egg mixture.

4. Tilt the pan to spread the egg mixture.

5. Cook until the underside is golden brown. Turn the omelet over and cook the other side until golden brown.

6. Carefully remove the omelet onto a plate and serve.

7. Repeat steps 3 – 6 and make the other omelet.

Banana Nut Bread

Preparation time: 15 minutes

Cooking time: 60 – 90 minutes

Makes: 2 loaves

Ingredients:

For dry ingredients:

- 2 teaspoons ground cinnamon
- 2 teaspoons baking powder
- 2 teaspoons baking soda
- 3 cups flour
- 1 cup whole wheat flour
- 2 teaspoons baking powder

Other ingredients:

- 1 cup brown sugar
- 1 cup granulated sugar
- 2 eggs
- 2 teaspoons milk
- 1 cup chocolate chips
- 1 cup canna-butter
- 6 bananas, mashed
- 1 teaspoon vanilla extract
- 1 cup chopped walnuts

Directions:

1. Add canna-butter, brown sugar, and sugar to a large mixing bowl and beat with an electric hand mixer until creamy.

2. Add eggs at a time and beat well each time. Scrape the sides of the bowl, time to time.

3. Beat until the mixture is light and creamy. Set aside for a while.

4. To mix dry ingredients: Sift together flour, whole-wheat flour, baking powder, baking soda, and cinnamon.

5. Add bananas, milk, and vanilla into another bowl and beat until well combined. Pour into the mixing bowl.

6. Whisk until well combined.

7. Add the mixture of dry ingredients and mix until well combined and free from lumps.

8. Add the walnuts and chocolate chips and fold gently.

9. Grease 2 loaf pans with some cooking spray. Divide the batter among the loaf pans.

10. Place the baking pan in an oven that has been preheated to 325° F and bake for 1 - 1 ½ hours or until brown on top.

11. Cool on your countertop for 15 minutes.

12. Invert on to a cooling rack. Cool for some more time.

13. Slice and serve. Store in an airtight container in the refrigerator.

Berry and Banana Smoothie

Preparation time: 5 minutes

Cooking time: 0 minutes

Makes: 1 – 2 servings

Ingredients:

- ½ cup coconut milk or almond milk, unsweetened

- ½ medium bananas, peeled, sliced

- ¾ cup frozen berries of your choice

- ½ – 1 tablespoon canna-coconut oil or any other canna-cooking oil

- 2 teaspoons chia seeds or ½ tablespoon hemp seeds (optional)

- ¼ cup orange juice

Directions:

1. Place bananas, berries, and chia seeds in a blender.

2. Pour milk, canna – oil, and orange juice. Blitz for 30 – 40 seconds or until smooth and creamy.

3. Pour into 1 – 2 glasses. Serve immediately with crushed ice.

Green Juice

Preparation time: 10 minutes

Cooking time: 0 minutes

Makes: 2 servings

Ingredients:

- 10 handfuls spinach

- 14 large cannabis fan leaves

- 1 lemon, peeled

- 6 kale leaves, torn

- 1 cucumber, chopped into chunks

- 2 Fuji apples, chopped into chunks

Directions:

1. Juice together spinach, cannabis fan leaves, lemon, kale, cucumber, and apples in a juicer.

2. Pour into 2 glasses and serve.

Sweet n Spicy Juice

Preparation time: 10 minutes

Cooking time: 0 minutes

Makes: 2 servings

Ingredients:

- 6 handfuls spinach

- 40 small sugar leaves

- 20 large cannabis fan leaves

- 1 jalapeño, chopped

- 10 kale leaves, torn

- 2 cucumbers, chopped into chunks

- 2 cups pineapple chunks

- 2 large cannabis buds

Directions:

1. Juice together spinach, sugar leaves, cannabis fan leaves, jalapeño, kale, cucumber, cannabis buds, and pineapple in a juicer.

2. Pour into 2 glasses. Add water to dilute if desired and serve with crushed ice.

Strawberry Banana Smoothie

Preparation time: 5 minutes

Cooking time: 0 minutes

Makes: 2 servings

Ingredients:

- 6 tablespoons canna-butter

- 1 cup frozen strawberries

- 1 cup thin vanilla yogurt

- Ice cubes, as required

- 2 bananas, sliced, frozen

- 6 tablespoons shredded coconut

- ½ cup coconut milk or milk of your choice

Directions:

1. Add canna-butter, strawberries, yogurt, ice cubes, bananas, coconut, and milk into a blender.

2. Blitz for 30 – 40 seconds or until smooth and creamy.

3. Pour into 2 glasses and serve

Raw Cannabis Green Smoothie

Preparation time: 10 minutes

Cooking time: 0 minutes

Makes: 2 – 3 servings

Ingredients:

- 10 handfuls spinach

- 6 kale leaves

- 14 large cannabis fan leaves

- 1 cucumber, chopped

- 2 Fuji apples, peel if desired, cored, chopped

- Juice of a lemon

Directions:

1. Add spinach, kale, cannabis, cucumber, apples, and lemon juice into a blender and blitz until smooth.

2. Pour into 2 – 3 glasses and serve.

Weed Iced Coffee

Preparation time: 2 minutes

Cooking time: 1 minute

Makes: 4 servings

Ingredients:

- 1 cup water

- 2 cups ice cubes or more if desired

- 1 cup canna- milk

- 4 teaspoons instant coffee granules

- 1 can (5 ounces) sweetened condensed milk

- 2 tablespoons chocolate syrup

Directions:

1. Pour water into a saucepan. Heat the water until warm. Turn off the heat.

2. Add instant coffee and stir until it dissolves.

3. Pour into a blender. Add ice cubes, canna milk, chocolate syrup, and condensed milk and blitz until well combined.

4. Pour into 4 glasses and serve.

Indian Bhang

Preparation time: 5 minutes

Cooking time: 2 – 3 minutes

Makes: 4 servings

Ingredients:

- ¼ ounce butter

- 0.07 ounce cannabis or hash

- 4 cups milk

- ¼ teaspoon ground cinnamon or ground nutmeg

- Honey or sugar to taste (optional)

Directions:

1. Add butter into a saucepan. Place saucepan over low heat. Add cannabis and let it cook for a minute in the melted butter.

2. Add milk and the chosen spice and stir. When milk is warm, turn off the heat.

3. Add honey or sugar to taste if desired.

4. You can serve as it is or with vodka.

CHAPTER FIVE

DINNER RECIPES

Main Course

Cannabis-Infused Butter Chicken

Preparation time: 4 hours

Cooking time: 30 minutes

Makes: 3 servings

Ingredients:

To marinate chicken:

- 1 pound boneless, skinless chicken breasts, cubed

- ¼ teaspoon freshly cracked pepper

- ½ teaspoon turmeric powder

- ¼ teaspoon sea salt or to taste

- ½ teaspoon chili powder

- ¼ cup Greek yogurt

For butter gravy:

- 2 tablespoons canna-butter

- 1 tablespoon butter

- 1 large onion, diced

- ½ teaspoon chili powder

- ½ teaspoon turmeric powder

- ½ tablespoon garam masala

- ½ teaspoon ground cumin

- ½ teaspoon cayenne pepper

- ¼ teaspoon pepper

- 2 cloves garlic, peeled, minced

- ½ tablespoon brown sugar

- ¼ cup water

- Salt to taste

To serve:

- A handful fresh cilantro, chopped

- Cooked, long grain rice

Directions:

1. To marinate chicken: Add yogurt, salt, and all the spices into a bowl and stir well.

2. Add chicken and stir until chicken is well coated with the marinade.

3. Cover and chill for 4 hours or longer if you have time on hand.

4. Place a skillet over medium flame. Add butter and let it melt.

5. Remove only chicken from the marinade and shake to drop off excess marinade. Place the chicken in the skillet and cook until brown all over.

6. Stir in the marinade and transfer into a bowl.

7. To make sauce: Add a tablespoon of canna-butter into the skillet.

8. When butter melts, add onion and sauté for a couple of minutes.

9. Stir in the ginger and garlic and sauté for about a minute until you get a nice aroma.

10. Stir in all the spices, salt, and brown sugar and cook for a few seconds.

11. Stir in tomato sauce and water and let it come to a boil.

12. Now add cream and stir. Let it come to a boil. Add chicken back into the pot along with the drippings and cook for 20 minutes or until chicken is cooked through.

13. Sprinkle salt and pepper to taste. Add a tablespoon of canna-butter and stir.

14. Sprinkle cilantro on top. Serve butter chicken over rice.

Cannabis Chicken Fajitas

Preparation time: 15 minutes

Cooking time: 30 minutes

Makes: 4 servings

Ingredients:

- 1 tablespoon canna-oil

- ½ teaspoon ground cumin

- ½ pound skinless, boneless chicken breasts

- 1 bell pepper, sliced

- 1 clove garlic, finely chopped

- 4 corn tortillas

- 3 tablespoons crumbled cotija cheese

- ½ teaspoon chili powder

- Freshly ground pepper to taste

- 1 tablespoon extra-virgin olive oil

- ½ red onion, sliced

- ¼ teaspoon grated lime zest

- Lime juice to taste

- ¼ cup prepared Pico de Gallo

Directions:

1. Combine ¼ teaspoon cumin, pepper, and salt to taste in a bowl. Sprinkle this mixture over the chicken and rub it well into it.

2. Place a large skillet over medium-high flame. Add ½ tablespoon extra-virgin olive oil. Once oil is hot, add chicken and cook until golden brown.

3. Remove chicken from the pan and place on a baking sheet.

4. Place the baking sheet into an oven than has been preheated to 350° F and bake for 10 - 15 minutes or until well-cooked.

5. Remove the chicken from the baking sheet and place on your cutting board. When cool enough to handle, cut into slices and keep warm.

6. Add ½ tablespoon extra-virgin olive oil into the skillet. Stir in onion, garlic, bell pepper, and cook for a couple of minutes.

7. Add remaining cumin and cook until vegetables are light brown.

8. Add canna-oil, lime zest, and a sprinkle of water. Add salt to taste. Cook for a couple of minutes and turn off the heat.

9. Heat the tortillas according to the instructions on the package.

10. Divide the chicken slices over the tortillas. Divide the vegetable mixture, cheese, and Pico de Gallo over the vegetables. Drizzle lime juice on top and serve.

Easy Cannabis-Infused Chicken Adobo

Preparation time: 15 minutes + marinating time

Cooking time: 30 minutes

Makes: 3 servings

Ingredients:

For marinade:

- 1 pound fresh chicken legs
- 1 teaspoon garlic powder
- 2 tablespoons soy sauce
- ½ teaspoon grated, fresh ginger
- 1 teaspoon freshly ground black pepper
- 1 tablespoon coconut oil

For sauce:

- 2 tablespoons canna-coconut oil
- 2 small bay leaves
- ½ medium onion, sliced
- 2 tablespoons apple cider vinegar
- ½ teaspoon maple syrup
- ¾ cup water
- 1 teaspoon whole peppercorns
- 2 cloves garlic, crushed
- 1 teaspoon brown sugar or to taste

- ½ teaspoon sea salt

Directions:

1. To marinate the chicken: Add soy sauce, garlic powder, ginger, and pepper into a bowl and stir.

2. Add chicken and stir until chicken is well coated with the marinade.

3. Cover and place it in the refrigerator for 1 – 8 hours.

4. To cook chicken: Place a pan over medium flame. Add oil. When oil melts and is well heated, place the chicken in the pan without the marinade.

5. Cook until brown all over.

6. To make sauce: Now add the marinade and water and let it come to a boil.

7. Add onion, garlic, peppercorns, and bay leaves and stir.

8. Cook on low until chicken is cooked through.

9. Stir in sugar, vinegar, salt, and maple syrup. Mix well and continue simmering for 5 – 8 minutes or until thick and you can see some oil. Stir every 3 – 4 minutes.

Chicken Pot-Cacciatore

Preparation time: 15 minutes

Cooking time: 50 – 60 minutes

Makes: 8 – 10 servings

Ingredients:

- 2 fryer chickens, with skin, cut into pieces, rinsed,

- 2 tablespoons canna-butter

- 2 tablespoons olive oil

- Pepper to taste

- 1 glass white wine (optional)

- ¼ cup whole green olives

- ¼ cup whole black olives

- Salt to taste

- 2 large onions, cut into ½ inch wedges

- 2 – 3 cups small cremini mushrooms

Directions:

1. Pat the chicken with paper towels until absolutely dry.

2. Season the chicken pieces with salt and pepper.

3. Place a large skillet over medium flame. Add oil as well as canna-butter.

4. Add chicken pieces once the butter melts and cook until brown all over.

5. Take out the chicken using a slotted spoon and place on a plate lined with paper towels. Do not discard the fat.

6. Add onions into the same pan and cook until slightly soft.

7. Add chicken and mushrooms and cook for 4-5 minutes.

8. Stir in wine and simmer for a couple of minutes. Add green olives and black olives and mix well. Turn off the heat. Cover and set aside for a few minutes before serving.

Cannabis-Infused Thai Green Curry

Preparation time: 15 minutes

Cooking time: 20 – 25 minutes

Makes: 2 – 3 servings

Ingredients:

For green curry paste:

- 3 green chilies, deseeded, chopped

- ½ inch ginger, peeled, grated

- ½ bunch coriander (roots, stalks, and leaves), rinsed well, chopped

- Juice of ½ lime

- Zest of ½ lime grated

- ½ inch galangal, peeled, chopped

- 1 shallot, roughly chopped

- 1 clove garlic, crushed

- 1 stalk fresh lemongrass, chopped

- 4 kaffir lime leaves or use extra lemon zest

- ½ tablespoon coriander seeds, crushed

- ½ teaspoon whole peppercorns, crushed

- ½ teaspoon ground cumin

- 1 teaspoon Thai fish sauce

- 1 ½ tablespoons olive oil

For curry:

- 2 medium potatoes, peeled, chopped into chunks

- ½ cup cubed zucchini

- 2 small cloves garlic, peeled, minced

- ½ cup diced red bell pepper

- 7 – 8 green beans or snap peas, trimmed, halved

- ½ tablespoon olive oil

- Thai green curry paste to suit your taste

- 1 cup coconut milk

- 1 cup chicken broth

- ½ teaspoon brown or cane sugar

- ½ pound boneless chicken, cut into bite-size pieces

- 1 kaffir lime leaf, thinly sliced or 2 strips lemon zest

- ½ Thai chili, sliced

- 2 teaspoons canna-butter

- ½ teaspoon vegetable oil

- Steamed rice to serve

Directions:

1. To make green curry paste: Add chilies, ginger, galangal, cilantro, garlic, lemongrass, shallot, lemongrass, kaffir leaves, spices, fish sauce, and oil into a blender and blend until smooth. You can also pound in a mortar and pestle.

2. To make curry: Cook potatoes in a pot of boiling water for 5 minutes or until fork-tender. Add zucchini, beans, and bell pepper.

3. Drain off after 3 minutes and set it aside.

4. Place a large pan over medium flame. Add oil and let it heat. Once oil is heated, add garlic and stir until it turns light golden brown in color.

5. Add curry paste, 2 – 4 teaspoons, or more if desired and stir-fry for about 20 seconds or until you get a nice aroma.

6. Add coconut milk and canna-butter and let it melt.

7. Add broth and stir. Add more broth if you want more gravy.

8. Add fish sauce, brown sugar, Thai chili, and chicken and stir. Lower the heat and cover with a lid. Simmer until chicken is cooked through.

9. Add cooked vegetables and heat thoroughly.

10. Add basil and turn off the heat. Garnish with kaffir leaves and serve over rice.

Classic Cannabis Lasagna

Preparation time: 20 minutes

Cooking time: 45 – 50 minutes

Makes: 2 – 3 servings

Ingredients:

- ½ pound ground turkey or beef

- ½ medium onion, finely chopped

- ½ can (from a 14.5 ounces can) stewed tomatoes

- ½ can (from a 6 ounces can) tomato paste

- 1 large egg, beaten

- ¼ cup ricotta cheese

- 1 teaspoon chopped fresh parsley

- ½ teaspoon pepper or to taste

- 4 ounces shredded cheddar cheese

- 1 ½ tablespoons canna-extra-virgin olive oil

- 1 clove garlic, minced

- ½ jar (from a 6 ounce jar) tomato sauce

- ½ box (from an 8 ounces bag) no-boil lasagna noodles

- ¾ cup cottage cheese

- 4 ounces shredded mozzarella cheese

- 4 ounces grated parmesan cheese

- 1 teaspoon salt

Directions:

1. Take rectangular baking dish and spray some oil in the dish.

2. Place a skillet over medium-low heat. Add canna-oil and heat slightly, making sure not to let the oil smoke.

3. Add garlic and onion and sauté for a minute. Add ½ teaspoon salt and ¼ teaspoon pepper. Mix well.

4. Stir in turkey and cook until brown. As the turkey is cooking, break it using a wooden spoon. Discard excess fat from the pan.

5. Add tomato sauce, stewed tomatoes, and tomato paste into the pan and stir. Cover and cook on low for about 10 minutes. Stir every 4 – 5 minutes.

6. Add egg, cottage cheese, ¼ cup parmesan cheese, ricotta cheese, parsley, ½ teaspoon salt, and ¼ teaspoon pepper into a bowl and mix until well incorporated.

7. Add a little of the turkey sauce into the baking dish and spread it in a thin layer.

8. Place a layer of lasagna noodles in the baking dish, overlapping by about ½ an inch.

9. Spread half the egg- cheese mixture over the noodles, followed by half the remaining mozzarella and half the cheddar cheese. Spread some more turkey sauce over this layer.

10. Repeat a layer of lasagna noodles. Spread remaining half egg-cheese mixture over the noodles, followed by half the remaining mozzarella and half the cheddar cheese. Spread some more turkey sauce over this layer.

11. Top with remaining parmesan cheese.

12. Place the baking dish in an oven that has been preheated to 350° F and bake for about 25-30 minutes or until you can see the sauce bubbling.

13. Fresh bread tastes great with this lasagna.

Cannabis-Infused Turkey Bolognese

Preparation time: 15 minutes

Cooking time: 30 minutes

Makes: 3 – 4 servings

Ingredients:

- 4-6 drops cannabis tincture

- Pepper to taste

- ½ cup grated parmesan cheese

- Salt to taste

- 2 tablespoons chopped parsley

- 2 cloves garlic, peeled, minced

- 1 stalk celery, finely chopped

- 1 medium onion, finely chopped

- 2 medium carrots, finely chopped

- ½ can (from a 7.2 ounces can) tomato sauce

- ½ can (from a 28 ounces can) chopped tomatoes

- ½ box whole wheat spaghetti

- ½ pound ground turkey

- 1 tablespoon olive oil

Directions:

1. Place a pot over medium-high flame. Once the oil is heated, add onion, carrot, and celery and stir. Cook until vegetables are tender.

2. Stir in the ground turkey and garlic and cook until brown, breaking it simultaneously as it cooks.

3. Stir in the tomato sauce and diced tomatoes.

4. While the turkey is cooking, cook spaghetti following the directions on the package. Drain but retain a little of the cooked water, about 3-4 tablespoons.

5. Toss together spaghetti and cannabis tincture and add into the pot. Add a little of the pasta cooked water and mix well.

6. Garnish with parsley and cheese and serve.

Beef and Bean Marijuana Chili

Preparation time: 20 minutes

Cooking time: 4 – 6 hours

Makes: 8 servings

Ingredients:

- 4 ounces dried New Mexico chili flakes (12 – 16 chilies)

- 1 tablespoon ground cumin

- 2 tablespoons salt or to taste

- 2 teaspoons dried oregano

- 2 tablespoons olive oil, divided

- 2 tablespoons canna- oil

- 1 ¾ cups beef stock, divided

- 3 medium onions, chopped

- 10-12 cloves garlic, peeled, minced (about 2 tablespoons)

- 2 tablespoons dark brown sugar

- 2 corn tortillas

- 4 teaspoons freshly ground pepper

- 1 teaspoon cayenne pepper

- 6 pounds boneless, chuck beef, trimmed of fat cubed (¾ inch cubes)

- 1 can (15 ounces) black beans, drained

- 1 can (15 ounces) dark red kidney beans, drained

- 1 cup black coffee

- 4 tablespoons apple cider vinegar

Directions:

1. Place a cast-iron skillet over medium-low flame. When the pan is heated, add chilies into the pan and cook until toasted. You should get an aroma. You need to keep a watch over it as it can get burnt. You need to stir often.

2. Remove the chilies from the pan and place them in a bowl. Pour hot water over the chilies and let it sit for about 30 minutes. Turn the chilies a couple of times.

3. Place the corn tortillas in the skillet one at a time and cook for a minute on either side.

4. Tear up the tortillas and add into a blender. Blend until smooth. Transfer into a bowl.

5. Add chilies (not the soaked water) into a blender. Deseed the chilies if desired and discard the stem.

6. Add ¾ cup beef stock, cumin, cayenne, pepper, canna-oil, oregano, and salt into the blender and blend until smooth.

7. Pour the blended mixture into a slow cooker. Set the slow cooker on high.

8. Add 1 ¼ tablespoon olive oil into the skillet. When the oil is heated, add beef in batches and cook until brown all over. Remove beef with a slotted spoon and place in a bowl.

9. When all of the beef is browned, discard the fat and add the beef into the slow cooker.

10. Add remaining oil into the skillet. When the oil is heated, add onion and garlic and stir-fry until light brown.

11. Transfer the onion mixture into the slow cooker. Also, add the blended tortillas, 1 cup beef stock, vinegar, both the variety of beans, coffee, and brown sugar and stir.

12. Keep the slow cooker covered and set the timer for 4 – 6 hours or until meat is cooked.

13. Ladle into bowls and serve.

Canna-Butter Pan-Seared Steak

Preparation time: 10 minutes

Cooking time: 5 minutes

Makes: 4 – 5 servings

Ingredients:

- 3 pounds ribeye steak, 1 ½ inches thick, at room temperature

- 1 teaspoon salt

- 2 teaspoons canna-butter

- 2 tablespoons olive oil

- Freshly ground black pepper to taste

Directions:

1. Place a large, ovenproof pan over high flame for about 4 – 5 minutes.

2. Brush oil all over the steak and season with salt and pepper.

3. Place steaks in the pan and cook for about half a minute, undisturbed.

4. Spread a little of canna butter over the steaks. Turn the steaks over and cook for half a minute. Spread a little of the canna butter over the steaks. Turn off the heat.

5. Shift the pan into an oven that has been preheated to 350° F and bake for 2 minutes on each side for medium-rare or 3 minutes on each side for medium cooked or 4 minutes for each side for well cooked.

6. Once the steaks are cooked to the desired doneness, take out the steaks from the pan and place on a serving platter. Tent loosely with aluminum foil. Let it sit for 2 minutes.

7. Serve.

Canna-Burgers

Preparation time: 10 minutes

Cooking time: 30 minutes

Makes: 6 – 7 servings

Ingredients:

- 2.2 pounds ground beef or its vegan equivalent

- 0.035 ounce cannabis, first decarbed and then ground into fine powder

- 2 onions, finely chopped

- 2 eggs, beaten

- Salt to taste

- 3 – 4 cloves garlic, peeled, minced

- Pepper to taste

- 1 teaspoon paprika

To serve:

- Tomato slices

- Cucumber slices

- Mayonnaise

- Burger buns

- Lettuce leaves

- Cheese slices etc.

Directions:

1. Add meat, cannabis, onion, eggs, salt, pepper, and cayenne pepper into a bowl.

2. Mix well using your hands. Dip your hands in water before mixing.

3. Divide the mixture into 6 – 7 equal portions and shape into burgers.

4. Place a pan over high flame. Spray some cooking spray in the pan.

5. Cook burgers in the pan for a minute on each side.

6. Now lower the flame to medium. Continue cooking on both the sides until the burgers are cooked to the desired doneness. You can also grill the burgers on a preheated grill.

7. Remove the burgers and place on a plate.

8. If you are using cheese, place the cheese slices on the burgers during the last minute of cooking.

9. Serve with any of the suggested serving options.

Weed-Infused Pulled Pork

Preparation time: 10 minutes

Cooking time: 1 ½ - 2 hours

Makes: 4 – 5 servings

Ingredients:

For pork:

- 2 ½ pounds boneless pork
- 1 ½ tablespoons canna-olive oil or canna-butter
- ½ teaspoon garlic powder
- ½ teaspoon ground cumin
- 6 ounces beer
- 1 ½ tablespoons brown sugar
- ½ tablespoon Himalayan salt
- ½ teaspoon smoked paprika
- Pepper to taste

For sauce:

- 14 tablespoons ketchup
- ¼ cup Dijon mustard
- 1 tablespoon Worcestershire sauce
- 6 tablespoons apple cider
- 2 tablespoons packed dark brown sugar
- 4 – 5 hamburger buns, split

Directions:

1. Trim the fat from the pork and cut into big pieces.

2. Combine canna-oil and spices in a baking dish. Place pork in the dish, and stir pork is until well coated with the mixture.

3. Place the baking dish in an oven that has been preheated to 400° F and bake for about 15 minutes.

4. Pour beer over and around the meat and continue roasting about 2 hours or until the meat is very well cooked and is breaking off.

5. Remove the baking dish from the oven and place it on your countertop.

6. To make sauce: Add ketchup, mustard Worcestershire sauce, apple cider, and dark brown sugar into a bowl and whisk well.

7. When the meat is cool enough to handle, shred the meat with a pair of forks into about 1-½ inch pieces.

8. Add meat into a pot. Add sauce and stir. Heat thoroughly.

9. Toast the buns if desired. Place pulled pork over bottom half of the buns. Cover with top half of the buns and serve.

BBQ Pork Ribs

Preparation time: 30 minutes

Cooking time: About 2 hours

Makes: 4 – 6 servings

Ingredients:

- 2 pounds baby back ribs, remove the tough membrane

- 1 teaspoon onion powder

- 1 teaspoon chili powder

- 1 tablespoon brown sugar

- 1 teaspoon garlic powder

- 1 teaspoon salt

- 1 teaspoon Hungarian paprika powder

- 1 cup cannabis-infused BBQ sauce

Directions:

1. Line a baking sheet with aluminum foil and place meat on it.

2. Add all the spices and salt into a bowl and stir. Sprinkle this mixture all over the meat. Keep the meat along with the baking sheet covered.

3. Place the baking dish in an oven that has been preheated to 350° F and bake for about 1 hour.

4. Uncover and spread cannabis-infused BBQ sauce over the meat. Continue baking for another 30 to 45 minutes.

5. Take out the baking dish from the oven and let the meat sit for 10 minutes.

6. Slice and serve.

Cannabis Garlic and Rosemary Pork Chops

Preparation time: 5 minutes

Cooking time: 20 – 30 minutes

Makes: 2 servings

Ingredients:

- 2 pork chops

- Freshly ground pepper to taste

- 1 clove garlic, minced

- ½ tablespoon canna-oil

- Salt to taste

- ½ tablespoon minced fresh rosemary

- 4 tablespoons canna-butter

Directions:

1. Sprinkle salt and pepper over the pork chops.

2. Add canna-butter, garlic, and rosemary into a bowl. Stir and keep it aside.

3. Place an ovenproof skillet over medium flame. Add canna-oil. When the oil is heated, place pork chops in the pan and cook until golden brown on both the sides. It should take about 4 minutes on each side. Turn off the heat.

4. Brush canna-butter mixture all over the pork chops.

5. Place the skillet into an oven that has been preheated to 350° F and bake for 10 to 13 minutes, depending on how you like it cooked.

6. Divide into 2 plates and serve with the remaining canna-butter mixture.

Jambalaya

Preparation time: 15 minutes

Cooking time: 45 minutes

Makes: 8 – 10 servings

Ingredients:

- 6 tablespoons cannabis olive oil

- 2 medium yellow bell peppers, chopped

- 2 green bell peppers, chopped

- 2 medium onions, chopped

- 2 cans (14.5 ounces each) fire-roasted or regular chopped tomatoes, with its liquid

- 2 packages instant Jambalaya rice mix

- 2 packages (12 ounces each) Andouille sausage, cut into ¼ inch thick slices

- ½ cup chopped fresh parsley (optional)

- 1 1/3 cups water

- 2 pounds large shrimp, peeled, deveined

- 2 teaspoons cayenne pepper or to taste

- Salt to taste

Directions:

1. Place a Dutch oven over low flame. Add canna-oil and let it heat. When the oil is heated, add onion and the bell peppers and cook until slightly tender.

2. Add Jambalaya rice mix, water, and tomatoes with its liquid and stir.

3. Increase the heat to medium and let the mixture come to a boil.

4. Lower the heat once again and cook covered for about 20 minutes. Stir every 7 – 8 minutes.

5. Add shrimp and sausage and keep the pot covered. Continue cooking until the rice is cooked. Stir in cayenne pepper and salt and turn off the heat.

6. Let the pot covered for 10 minutes.

7. Garnish with parsley and serve with rice or pasta.

Baked Shrimp Scampi

Preparation time: 15 minutes

Cooking time: 12 minutes

Makes: 3 servings

Ingredients:

- 1 pound shrimp in shell, peeled, deveined, keep the tails

- 1 tablespoon dry white wine

- 6 tablespoons canna-butter, at room temperature

- 1/8 cup minced shallots

- ½ teaspoon minced fresh rosemary leaves

- ½ teaspoon grated lemon zest

- Yolk of 1 medium egg

- Lemon wedges to serve

- 1 ½ tablespoons canna-oil

- Kosher salt to taste

- 2 teaspoons minced garlic

- 1 ½ tablespoons minced fresh parsley

- 1/8 teaspoon crushed red pepper flakes

- 1 tablespoon fresh lemon juice

- 1/3 cup panko bread crumbs

- Pepper to taste

Directions:

1. Add shrimp into a bowl. Pour wine and oil over it. Sprinkle salt and pepper to taste. Toss well. Set aside for 10 minutes.

2. Add garlic, shallots, herbs, red pepper flakes, butter, lemon juice, lemon zest, panko breadcrumbs, yolk, salt, and pepper into a bowl and mix well.

3. Take a small oval dish of about 9-10 inches. Place shrimp on the bottom of the dish, with the curled tail side facing up. Place them in a single layer.

4. Spoon the breadcrumb mixture evenly over the shrimp.

5. Bake in a preheated oven at 425° F for about 10 -12 minutes or until the mixture is bubbling.

6. For a brown top, broil for a minute.

7. Let it cool for a couple of minutes before serving. Top with lemon wedges and serve.

Grilled Fish Tacos with Ganja Green Salsa

Preparation time: 15 minutes

Cooking time: 15 minutes

Makes: 4 servings

Ingredients:

For salsa:

- 2 large green chilies like Hatch chilies
- 2 tomatillos, remove the husk, halved
- 1 teaspoon minced garlic
- 2 tablespoons canna-oil
- 2 jalapeño peppers
- 1 small onion, quartered
- 2/3 cup minced cilantro
- Lime juice to taste

For fish:

- 2 teaspoons vegetable oil
- 2 pounds mahi-mahi or other firm fish
- Salt to taste
- 2 cups shredded green cabbage
- 2/3 cup crumbled queso fresco
- Pepper to taste
- 2 medium avocadoes, peeled, pitted, diced

- 8 large corn tortillas

Directions:

1. Preheat a grill and place chilies, tomatillos, jalapeños, and onion pieces on the grill and grill until charred. Turn the vegetables on the grill so that it is evenly charred.

2. Remove the vegetables from the grill. Place the chilies and jalapeños in a paper bag and close the bag. Let it sit for 10 minutes.

3. Peel the skin from the chilies and jalapeños and place them in a blender. Also, add cilantro, garlic, lime juice, and canna-oil and blend until pureed.

4. Add salt and pepper and stir. Transfer into a bowl. Cover and set aside.

5. To make fish: Brush oil all over the fish. Sprinkle salt and pepper all over the fish.

6. Set up your grill and preheat it to medium heat. Place fish on the grill and grill for 2 to 3 minutes on each side or until it flakes easily when pierced with a fork.

7. Warm the tortillas following the instructions on the package.

8. Scatter cabbage on the tortillas. Place fish on each tortilla. Drizzle some chili salsa over it. Scatter avocado and cheese on each tortilla and serve.

Weed Fish and Chips

Preparation time: 10 minutes

Cooking time: 10 minutes

Makes: 2 servings

Ingredients:

- 2 large potatoes, peeled, cut into fries

- ½ teaspoon baking powder

- ½ teaspoon pepper

- 1 small egg

- ¾ pound cod fillets

- ½ cup all-purpose flour

- ½ teaspoon salt

- ½ cup canna-milk

- Oil to fry, as required

Directions:

1. Immerse potatoes in a bowl of cold water.

2. Add flour, salt, baking powder, and pepper into a bowl and stir well.

3. Add milk and egg and whisk well. Leave it aside to rest for about 20 minutes.

4. Meanwhile, place a deep fryer pan over medium heat. Pour enough oil to fill the pan (about 2-3 inches). Let the oil heat to 350° F.

5. Add potatoes in batches into the pan and cook until fork tender.

6. Remove potatoes with a slotted spoon and place on a plate lined with paper towels.

7. Dip the fish in the batter, one at a time, and carefully drop it in the hot oil. Cook until golden brown all over.

8. Remove fish with a slotted spoon and place on a plate lined with paper towels.

9. Once you are done with the frying of fish, add the potatoes back into the hot oil and cook for a couple of minutes until crisp.

10. Serve fish with chips.

Sativa Shrimp Creole

Preparation time: 15 minutes

Cooking time: 30 minutes

Makes: 2 servings

Ingredients:

- ½ tablespoon unsalted butter

- 1 tablespoon all-purpose flour

- ¼ green bell pepper, diced

- 1 teaspoon minced garlic

- ¼ small onion, diced

- ½ stalk celery, finely diced

- ½ can (from a 15 ounces can) crushed tomatoes with its juices

- 2 small bay leaves

- 1 tablespoon chopped fresh Italian parsley or 1 teaspoon dried parsley

- 0.004 ounce decarbed kief or finely ground hash

- Cooked rice to serve

- ½ tablespoon olive oil

- ½ cup stock

- A pinch cayenne pepper or to taste

- Salt to taste

- ½ pound medium shrimp, peeled

Directions:

1. Place a skillet over medium flame. Add butter and olive oil and let the butter melt.

2. Add flour and stir continuously until roux is formed and is light brown in color.

3. Stir in celery, bell pepper, and onion and keep stirring until the vegetables are slightly tender.

4. Add garlic and cook for a few seconds until you get a nice aroma.

5. Add tomatoes with its juices, bay leaves, parsley, kief, stock, cayenne pepper, and salt, then stir.

6. Bring it to a boil over high heat. Now lower the heat to low heat and cook for about 10 minutes.

7. Stir in shrimp and cook until pink.

8. Divide rice into plates. Divide shrimp creole and spoon over the rice. Serve immediately.

Butter Garlic Shrimp

Preparation time: 10 minutes

Cooking time: 15 minutes

Makes: 8 servings

Ingredients:

- 2 pounds shrimp, peeled, deveined

- 1 cup canna-butter

- ½ teaspoon red pepper flakes

- 2/3 cup chopped parsley

- Angel hair pasta, to serve (optional)

- 4 tablespoons canna-olive oil

- 12 cloves garlic, peeled, minced

- 6 tablespoons lemon juice

- Salt to taste

Directions:

1. Place a large skillet over medium flame. Add canna-olive oil and let it heat. Add shrimp into the pan and spread it evenly. Do not stir for 3 – 4 minutes.

2. Sprinkle salt over it and stir. When shrimp begins to become light pink, stir in garlic and red pepper flakes.

3. Sauté for a couple of minutes. Stir in lemon juice, ½ cup canna-butter, and 1/3-cup parsley. When butter melts, lower the heat and add remaining butter.

4. Let it simmer until slightly thick.

5. Remove shrimp with a slotted spoon and add into a bowl. Add 1/3-cup parsley and toss well.

6. Let the sauce simmer for 3 – 4 minutes. If the consistency of the sauce is thick, add water to dilute, using 1 teaspoon at a time, stirring each time.

7. Turn off the heat. Add salt to taste and stir.

8. To serve as main course, serve over angel hair pasta. Drizzle sauce on top and serve. You can also serve it with steak.

9. To serve as appetizer, serve shrimp on a serving platter with sauce in a bowl as a dip.

Cannabis-Infused Pasta with Clams and Green Chiles

Preparation time: 15 minutes

Cooking time: 30 minutes

Makes: 2 – 3 servings

Ingredients:

- ¼ cup + ½ tablespoon extra-virgin olive oil

- ¼ cup lightly packed parsley leaves

- 1 small poblano pepper, discard stem, deseeded, chopped

- 1 ounce shishito peppers, discard stems, deseeded, chopped

- Kosher salt to taste

- 1 shallot, thinly sliced

- 1 small shallot, minced

- 1 tablespoon drained capers

- ½ cup dry white wine

- 21 mixed clams like Manila, littleneck and razor, scrubbed

- 1 ½ tablespoons unsalted butter

- 2 tablespoons crème fraiche

- Wasabi caviar, to garnish

- A large handful mint leaves + extra to garnish

- 2 tablespoons snipped chives

- 1 Cubanelle pepper, discard stem, deseeded, chopped

- Pepper to taste

216

- 1 clove garlic, minced

- 1 clove garlic, crushed

- 1 teaspoon whole peppercorns

- 1 teaspoon fennel seeds

- 1 teaspoon coriander seeds

- 1 teaspoon mustard seeds

- 1 cup bottled clam juice

- 6 ounces pipe rigate or mezze rigatoni pasta

- ½ tablespoon canna-butter

Directions:

1. Add ¼ cup oil, parsley, half the mint leaves, and chives into a blender and blend until smooth.

2. Place a fine wire mesh strainer over a bowl. Pass the blended mixture through the strainer. Press to remove as much liquid as possible. Throw away the solids.

3. Place a cast-iron skillet over medium-high flame. Add ½ tablespoon oil and let it heat well, almost up to smoking point.

4. Place all the varieties of chilies in the pan. Season with salt and pepper and cook until charred with few blisters on the peppers. Stir occasionally.

5. Stir in the minced shallot and minced garlic. Add capers and cook on low for a couple of minutes and turn off the heat.

6. After about 10 minutes, add the strained oil into the pan and stir.

7. Place a pot over medium flame. Add whole spices into the pan and cook for 2 – 3 minutes, stirring frequently, until you get a nice aroma in the air.

8. Now stir in the sliced shallot and crushed garlic. Pour clam juice and stir.

9. When the juice comes to a boil, add clams, and cover the pot. Increase the heat to high heat and cook for about 8 – 9 minutes. By now, the clams should have opened up. Discard any clams that have not opened up.

10. Remove clams with a pair of tongs and place them on a baking sheet. When it cools a bit, take out the meat from the shells and throw off the shells.

11. Place a strainer over a bowl and pass the cooked liquid through the strainer. Use only the liquid and not the solids; they can be thrown away.

12. In the meantime, cook pasta, following the directions on the package. Drain and set aside.

13. Wipe the pot in which you have cooked pasta. Place the pot over medium-high flame.

14. Add butter. When butter melts, add clams, pasta, ¼ cup clam cooked liquid, chili mixture, and crème fraiche and stir. Heat thoroughly.

15. Add remaining mint leaves, canna-butter, lime juice, salt, and pepper and mix well. Turn off the heat.

16. Garnish with mint leaves and wasabi caviar and serve.

Weed Grilled Cheese

Preparation time: 5 minutes

Cooking time: 6 minutes

Makes: 2 servings

Ingredients:

- 2 slices cheddar cheese

- 4 teaspoons butter

- 4 teaspoons canna-butter

- 4 slices bread

Directions:

1. Apply a teaspoon of butter on one side of each of the bread slices.

2. Apply a teaspoon of canna-butter on the other side of each of the bread slices.

3. Lay the cheese slice over the canna-butter side, on 2 slices of bread. Cover with the remaining 2 slices of bread, with the canna-butter side facing down.

4. Place a nonstick pan over medium flame. When the pan is heated, place the sandwich in the pan and cook until the bottom side is golden brown. Turn the sandwich over and cook the other side until golden brown. Remove the sandwich and cut into the desired shape.

5. Cook the other sandwich similarly.

6. This tastes great with tomato soup.

Smoked Mac 'n' Cheese

Preparation time: 10 minutes

Cooking time: 50 – 60 minutes

Makes: 2 – 3 servings

Ingredients:

- ¼ cup unsalted butter
- ¼ cup cold canna- butter
- ½ tablespoon melted canna-butter
- 2 cups milk
- 1 teaspoon salt or to taste
- ½ teaspoon pepper or to taste
- ½ cup grated, smoked mozzarella cheese
- ½ cup grated parmesan cheese, divided
- ½ cup shredded cheddar cheese
- ½ cup shredded American of Swiss cheese
- ½ pound penne pasta
- ½ cup flour
- 1/8 teaspoon cayenne pepper or to taste
- 2 tablespoons breadcrumbs

Directions:

1. Cook pasta, following the directions on the package.

2. Place a skillet over medium flame. Add butter and cold canna-butter. When butter melts, stir in flour and cook for 2-3 minutes until roux is formed and cooked to the desired color.

3. Meanwhile, heat milk in a saucepan. When it nearly begins to boil, but not boiling, pour milk into the skillet, stirring constantly. Stir in salt, pepper, and cayenne pepper.

4. Cook until thick. Keep stirring until the mixture begins to boil. Turn off the heat. Add pasta, smoked mozzarella cheese, half the parmesan cheese, cheddar cheese, and American cheese. Mix well.

5. Spray a baking dish with some cooking spray.

6. Transfer the mixture into the baking dish.

7. Add breadcrumbs into a bowl. Add melted butter and mix well. Scatter this mixture over the pasta in the baking dish.

8. Place the baking dish in a preheated oven and bake at 400° F until golden brown on top.

9. Serve.

Marijuana Pizzadillas

Preparation time: 5 minutes

Cooking time: 1 minute

Makes: 2 servings

Ingredients:

- 4 small flour tortillas

- 2/3 cup pizza sauce

- ½ cup grated mozzarella cheese

- 1 teaspoon canna-oil

- ¼ - ½ cup pepperoni or cooked sausage (optional)

- ½ cup finely chopped vegetables of your choice (optional)

Directions:

1. Combine pizza sauce and canna-oil in a bowl.

2. Take 2 paper plates and place a tortilla on each. Scatter 2 tablespoons cheese on each tortilla.

3. Drop the pizza sauce in small amounts all over the tortillas. Scatter pepperoni and vegetables if using. Scatter remaining cheese all over the vegetables.

4. Cover with the remaining tortillas. Place the Pizzadillas in the microwave, one at a time, and cook for 1 minute each.

5. Remove from the microwave and let it cool for about a minute.

6. Cut into wedges and serve.

Cannabis-Infused French Bread Pizza

Preparation time: 5 minutes

Cooking time: 10 minutes

Makes: 2 servings

Ingredients:

For sauce:

- 1 ounce tomato paste

- ½ tablespoon canna-oil

- ¼ teaspoon minced garlic

- Crushed red pepper flakes to taste

- 1 ounce water

- ½ tablespoon olive oil

- ½ teaspoon Italian seasoning blend

- ¼ teaspoon balsamic vinegar

For pizza:

- 6 tablespoons shredded mozzarella cheese

- 1 small, soft loaf French or Italian bread, halved lengthwise

- 2 tablespoons grated parmesan cheese

- Toppings of your choice like onion, pepperoni, mushrooms, etc.

Directions:

1. To make pizza sauce: Combine tomato paste, canna-oil, garlic, red pepper flakes, water, olive oil, Italian seasoning, and vinegar in a bowl.

2. Place the bread halves on a plate, with the cut side facing up. Apply half the sauce on each bread. Scatter mozzarella on the bread slices. Next, scatter Parmesan, followed by any toppings.

3. Place rack in the center of the oven and preheat it to 400° F.

4. Place the bread pizzas on the rack and bake until cheese melts and is brown at a few spots.

5. Serve hot.

Fettuccine Alfredo Pasta

Preparation time: 15 minutes

Cooking time: 15 minutes

Makes: 3 servings

Ingredients:

- 12 ounces dry fettuccine pasta

- ¾ cup heavy cream

- ¼ cup grated parmesan cheese

- 1/3 heaping cup grated Romano cheese

- Garlic salt to taste

- Salt to taste (optional)

- ½ cup canna-butter

- Pepper to taste

Directions:

1. Cook fettuccini pasta in salted water, following the directions on the package.

2. Place a saucepan over low flame. Add canna-butter and let it melt.

3. Stir in cheese, salt, garlic salt, and pepper. Stir often until cheese melts.

4. Add cooked pasta and toss well.

5. Serve.

Side Dish

Mini Cannabis Green Bean Casserole

Preparation time: 15 minutes

Cooking time: 30 minutes

Makes: 8 servings

Ingredients:

- 6 cups fresh green beans, trimmed, chopped into 2 inch pieces

- 4 tablespoons butter

- 6 tablespoons canna-butter

- 1 large onion, chopped

- 2 cans (12.8 ounces each) cream of mushroom soup

- Salt to taste

- 2 cups sliced mushrooms

- 3 cups French fried onions, divided

- Pepper to taste

Directions:

1. Place a pot of water over high heat. When the water begins to boil, add green beans and cook for a few minutes until crisp as well as tender.

2. Drain in a colander.

3. In the meantime, place a large skillet over medium heat. Add canna-butter.

4. When the butter melts, add onion and mushrooms and sauté until tender. Turn off the heat.

5. Add beans, half the French fried onions, pepper, salt, and cream of mushroom soup and mix well.

6. To prepare ramekins and oven: Grease 8 ramekins with butter. You need to preheat the oven to 375° F.

7. Spoon the green bean mixture into the prepared ramekins. Scatter remaining French fried onions on top.

8. Bake until the top is golden brown.

Mashed Cannabis Cauliflower with Parmesan Cheese

Preparation time: 10 minutes

Cooking time: 10 minutes

Makes: 2 – 3 servings

Ingredients:

- 1 ¼ pounds cauliflower, cut into florets
- 3 tablespoons heavy whipping cream
- ¼ teaspoon pepper
- ½ cup shredded parmesan cheese, divided
- ½ tablespoon canna-butter
- 1 tablespoon minced parsley
- Salt to taste

Directions:

1. Place a pot half-filled with water over high heat. When the water begins to boil, add cauliflower florets and cook for a few minutes until tender.

2. Turn off the heat and keep the pot covered for about 15 minutes. Drain in a colander.

3. Transfer the cauliflower into a bowl and mash it with a potato masher. Add half the cheese, canna-butter, cream, salt, and pepper.

4. Garnish with rest of the cheese and parsley.

Cannabis Mashed Potatoes

Preparation time: 15 minutes

Cooking time: 20 minutes

Makes: 2 – 3 servings

Ingredients:

- 1 ¼ pounds russet potatoes, peeled, cut into 1 ½ inch cubes

- 2 teaspoons canna-butter

- 1 ounce cream cheese

- Kosher salt to taste

- Freshly ground pepper to taste

- 1 tablespoon butter, at room temperature

- Chopped chives to garnish

Directions:

1. Add water to a pot and place the pot over medium heat.

2. Add potatoes and about a teaspoon of salt into it. Cook until the potatoes are soft.

3. Drain the potatoes and discard the water. Add the potatoes, canna-butter, butter, cream cheese, salt, and pepper into a bowl. Mash with an electric hand mixer until smooth, or the texture you prefer is achieved.

4. Garnish with chives and serve.

Caramelized Brussels Sprouts

Preparation time: 5 minutes

Cooking time: 20 minutes

Makes: 2 – 3 servings (4 Brussels sprouts per serving)

Ingredients:

- 1 slice uncured bacon, cut into pieces

- ½ pound fresh Brussels sprouts, halved

- 2 cloves garlic, minced

- 1 tablespoon light agave nectar or maple syrup

- 1 tablespoon canna-oil

- ½ small onion, thinly sliced

- Pepper to taste

- 2 tablespoons balsamic vinegar or apple cider vinegar

- Salt to taste

Directions:

1. Place a pan over medium flame. Add bacon and cook until slightly crisp. Remove bacon with a slotted spoon and place it on a plate lined with paper towels.

2. Add canna-oil into the pan. Once oil is heated, add onion and cook until light golden brown.

3. Stir in Brussels sprouts and garlic and until brown, stirring occasionally.

4. Stir in the vinegar, salt, pepper, and agave nectar. Cook until dry.

5. Transfer into a bowl and serve topped with bacon.

Canna-Butter Sautéed Mushrooms

Preparation time: 5 minutes

Cooking time: 5 – 7 minutes

Makes: 2 servings

Ingredients:

- 1 – 2 tablespoons canna-butter

- ½ pound mushrooms, halved or quartered or whole, according to the size

Directions:

1. Place a pan over medium flame. Add canna-butter. When butter melts, add mushrooms and sauté until tender.

2. Serve immediately.

Broccoli Cheddar Cannabis Casserole

Preparation time: 10 minutes

Cooking time: 45 minutes

Makes: 4 servings

Ingredients:

- 5 tablespoons canna-butter, melted

- 4 cups crushed Ritz crackers or any other crackers

- 2 cups mayonnaise

- 2 cups grated sharp cheddar cheese

- 2 ½ pounds broccoli, cut into bite-size florets

- 4 eggs, lightly beaten

- 2 cups condensed cream of mushroom soup

Directions:

1. Place a pot of water over high heat. When the water begins to boil, add broccoli and cook for a few minutes until crisp as well as tender. They will turn bright green in color.

2. Drain in a colander. Immediately immerse broccoli in a bowl of ice water. Drain after 5 minutes in a colander.

3. Add broccoli, eggs, mushroom soup, cheddar cheese, and mayonnaise into a bowl and mix until broccoli is well coated with the mixture.

4. You need to prepare a baking dish and oven: For the oven, preheat it to 375° F. Spray a baking dish with cooking spray.

5. Add the broccoli mixture into the casserole dish

6. Pour melted canna-butter all over the broccoli mixture and swirl the pan to spread the butter all over.

7. Place the casserole dish in the oven and bake for about 30 – 40 minutes or until light golden brown on top.

8. Serve.

Cannabis-Infused Radical Ratatouille

Preparation time: 15 minutes

Cooking time: 20 minutes

Makes: 3 servings

Ingredients:

- 1 ½ tablespoons olive oil
- ½ tablespoon minced garlic
- 1 small zucchini, 1 inch dice
- ½ bell pepper or any color, cut into 1 inch squares
- 1 cup cubed eggplant, 1 inch cubes
- ½ cup cubed yellow summer squash
- ¾ cup diced tomatoes
- ½ large onion, cut into 1 inch dice
- ¼ teaspoon dried thyme
- ½ teaspoon pepper
- ½ teaspoon dried oregano
- Salt to taste
- 1 tablespoon canna-oil
- A handful fresh cilantro, chopped

Directions:

1. Place a cast-iron skillet over medium-high flame. Add ¾ tablespoon oil and let it heat.

2. Add onion and cook until slightly tender. Add garlic and cook for a few seconds until fragrant.

3. Stir in remaining oil and eggplant and cook for 4 – 5 minutes. Stir in squash and zucchini. Stir often and cook until slightly tender.

4. Stir in tomatoes, dried herbs, pepper, and salt. Cook until vegetables are tender.

5. Add canna-oil and basil and mix well.

6. Serve.

Almond, Orange, and Cucumber Stuffed Avocado

Preparation time: 15 minutes

Cooking time: 10 minutes

Makes: 4 servings

Ingredients:

- 6 tablespoons balsamic vinegar

- 4 small Clementine oranges, peeled, separated into segments

- A handful fresh cilantro, chopped

- ½ teaspoon fresh lemon juice

- Salt to taste

- 2 large avocadoes, halved, pitted

- 2 small Persian cucumbers, peeled, diced

- 3 tablespoons sliced, toasted almonds

- Pepper to taste

- 1 ½ tablespoons canna-olive oil

Directions:

1. Pour vinegar into a small saucepan. Place the saucepan over medium flame. Cook until it is reduced to 3 tablespoons.

2. Scoop out the avocado and cut into small cubes. Leave the shells of the avocado with a little of the flesh on it so that it remains stable when you fill it up.

3. Add avocado into a bowl along with rest of the ingredients. Mix well and fill this mixture into the avocado halves.

4. Trickle reduced vinegar on top and serve.

Weed Bread

Preparation time: 15 – 20 minutes + rising time

Cooking time: 40 – 50 minutes

Makes: 2 loaves or 1 large loaf

Ingredients:

- 0.35 ounce marijuana flowers, deseeded, de-stemmed

- Flaxseed meal or cornmeal, to dust

- ½ teaspoon active dry yeast

- 6 cups all-purpose flour or mixture of 4 cups flour and 2 cups whole wheat flour

- 2 ½ teaspoons salt

- 3 ½ cups lukewarm water (110°F)

Directions:

1. Decarb the marijuana and grind it coarsely.

2. Combine marijuana flowers, yeast, flour, and salt in a mixing bowl. Pour lukewarm water and mix to form sticky dough.

3. Keep the bowl covered with cling wrap and place it in a warm area, say on top of your refrigerator, for 12-16 hours or until the dough doubles in size and tiny bubbles should be visible on the top of the dough.

4. Dust your countertop as well as your hands with cornmeal or flaxseed meal.

5. Place the dough on your countertop. Divide the dough into 2 equal portions. You can also make one large loaf. Fold each portion of the dough a few times to form into a ball. Dredge the dough in some more cornmeal.

6. Place dough on a cotton cloth, with its seam side facing down. Sprinkle some cornstarch on top of the dough.

7. Wrap the dough lightly with the cloth and set aside for 2 hours to rise.

8. You need to preheat the oven to 450° F for 30 minutes before baking. Place a large metal pan or Dutch oven in the oven while preheating.

9. Remove the dough from the cloth and place in the heated loaf pan carefully with its seam side facing up.

10. Cover with a lid of a dutch oven or aluminum foil, place it in the oven, and bake for 40-50 minutes.

11. Uncover and continue baking for another 8-10 minutes. This is done to get a crust on top.

12. Switch off the oven and let the bread remain in the oven for 5 minutes. When you tap the bread, you should hear a hollow sound if the bread is ready.

13. Carefully remove the bread from the pan and cool on a wire rack completely.

14. Slice and serve. Store leftovers in an airtight container.

Cannabis Gravy

Preparation time: 5 minutes

Cooking time: 10 minutes

Makes: 20 – 24 servings

Ingredients:

- 1 cup canna-butter

- Salt to taste

- 4 onions, thinly sliced

- 1 cup balsamic vinegar

- Chicken stock or turkey stock, as required

- Pepper to taste

- ¼ chopped cup fresh sage,

- ¼ chopped cup fresh rosemary

- 2/3 cup flour

Directions:

1. Place a large skillet over low flame. Add canna-butter.

2. When the butter just melts, add onions and sauté until translucent. Make sure that the flame is low all the time.

3. Add rosemary and sage and cook for another 10 minutes.

4. Add flour and sauté for about 60 to 90 seconds, stirring constantly.

5. Pour the stock, stirring simultaneously. Cook until the gravy thickens.

6. Add vinegar and simmer for just 15 minutes and not longer.

7. Add salt and pepper to taste.

Weed French Fries

Preparation time: 15 minutes

Cooking time: 20 minutes

Makes: 2 servings

Ingredients:

- 1/3 cup canna- extra-virgin olive oil

- 1 ½ tablespoons salt

- 2 large potatoes, peeled, cut into fries

Directions:

1. Place the potatoes on a baking sheet.

2. Sprinkle salt and oil all over them. Toss well. Spread it evenly all over the baking sheet without overlapping.

3. Place the baking sheet in an oven that has been preheated to 400° F and bake for 20 to 30 minutes, depending on the way you like it cooked.

4. Serve hot.

CHAPTER SIX

DESSERT RECIPES

Chocolate Pudding

Preparation time: 10 minutes

Cooking time: 15 minutes

Makes: 4 – 6 servings

Ingredients:

- 2/3 cup sugar

- 1/3 cup cocoa powder

- 2 cups milk

- 2 tablespoons grated chocolate

- 2 tablespoons canna-butter

- 3 tablespoons cornstarch

- ½ teaspoon vanilla extract

Directions:

1. Add cocoa, sugar, cornstarch, and salt into a small pot and whisk well.

2. Add milk and whisk until smooth.

3. Place the pot over medium flame and stir constantly until it comes to a boil. After about 30 seconds, turn off the heat.

4. Stir in vanilla and canna butter. Pour the mixture into pudding cups or bowls. Let it cool completely.

5. Chill until use. Garnish with grated chocolate and serve.

French Toast Cupcakes

Preparation time: 15 minutes

Cooking time: 25 minutes

Makes: 6 servings

Ingredients:

For cupcakes:

- ¾ cup all-purpose flour

- ¾ teaspoon baking powder

- ¼ teaspoon ground allspice

- ¼ teaspoon salt

- ½ cup sugar

- ½ teaspoon ground cinnamon

- 1/8 teaspoon freshly grated nutmeg

- ¼ teaspoon ground allspice

- ¼ cup canna-butter

- 1 large egg

- 2 slices cooked bacon, cut each into 3 pieces

- ¼ cup sour cream

For topping:

- 2 tablespoons all-purpose flour

- 2 tablespoons chopped pecans

- 1 ¼ tablespoons butter, cubed, chilled

- 2 tablespoons sugar

- ¼ teaspoon ground cinnamon

Directions:

1. To make topping: Add flour, butter, pecans, sugar, and cinnamon into a bowl and mix until the butter is pea-size. Keep the bowl in the refrigerator, covered.

2. To prepare oven and muffin pan: Take a 6 counts muffin pan and place disposable paper liners in them. Place rack in the center of the oven and preheat the oven to 350° F.

3. Combine flour, baking powder, salt, spices, and sugar in a bowl.

4. Add sour cream, canna-butter, maple extract, and egg in a bowl and beat with an electric hand mixer set on medium speed until creamy.

5. Lower the speed and add the mixture of dry ingredients. Continue beating until just incorporated, making sure not to overbeat.

6. Pour batter into the prepared muffin tins, up to ¾ the muffin cups.

7. Divide equally the topping and scatter it over the batter in the muffin cups.

8. Place the muffin pan in the oven and bake for about 25 minutes. When the muffins are ready, if you pierce a toothpick in the middle of the cupcake, it should come out without any particles stuck on it.

9. Take out the muffin pan from the oven and let it remain on your countertop for 15 minutes.

10. Remove the muffins from the pan and place it on a cooling rack. Insert a piece of bacon in each muffin and serve.

11. Store leftovers in an airtight container in the refrigerator.

Key Lime Kickers

Preparation time: 5 minutes

Cooking time: 20 minutes

Makes: 12 servings

Ingredients:

- 3 tablespoons heavy cream

- 2 tablespoons weed sugar

- 4-5 drops key lime oil

- 1 tablespoon unsalted butter

- 5 ounces white chocolate, coarsely chopped

- Graham cracker crumbs to coat

Directions:

1. To make ganache, set up double boiler: Take 2 pots of nearly the same (but not same) sizes such that the smaller one fits inside, the larger pot.; the smaller pot should not touch the bottom of the bigger pot. It should fit well inside it.

2. Pour enough water into the larger pot such that it is 1/3 full. The water should not touch the smaller pot. Place the bigger bowl over medium flame. Let the water come to a boil.

3. Add cream into the smaller pot. Place the smaller pot inside the bigger pot.

4. Lower heat to low heat and let the water simmer. Add corn syrup, butter, and weed sugar and stir.

5. Stir in white chocolate. Stir occasionally. When the mixture is smooth, remove the bowl from the double boiler.

6. Take a bowl to make the dessert and weigh the bowl. Note down the weight.

7. Pat the bowl dry from the sides and bottom of the pan. Pour into the weighed bowl.

8. Place this bowl in the freezer until set, but not hard. You should be able to shape the ganache. Now weigh the bowl once again.

9. Use this formula: Weight of the bowl with ganache – (minus) weight of the bowl without ganache divided by 24. This formula is to get the weight of a truffle.

10. Add cracker crumbs into a shallow bowl.

11. Scoop out ganache (equal to the weight got from the formula) and place it on a baking sheet lined with parchment paper.

12. Dredge the ganache in cracker crumbs and place on another baking sheet lined with parchment paper.

13. Cover the baking sheet and chill until use. It can last for 4 – 5 weeks.

Brownies

Preparation time: 10 – 12 minutes

Cooking time: 20 minutes

Makes: 9 servings

Ingredients:

- 2 tablespoons salted butter

- 6 tablespoons canna-butter

- 2 ounces unsweetened chocolate

- 5 ounces semi-sweet chocolate

- 3 tablespoons unsweetened cocoa powder

- 1 ¼ cups sugar

- ½ tablespoons pure vanilla extract

- 3 large eggs

- ½ teaspoon salt

- 1 cup all-purpose flour

For drizzle:

- 1/8 teaspoon vegetable shortening

- ¼ cup white chocolate chips

Directions:

1. To prepare baking dish and oven: Spray a square baking dish (8 – 9 inches) with cooking spray and set it aside. Place rack in the center of the oven and make sure that the oven is preheated to 350° F.

2. Place a saucepan over low flame. Add canna-butter and let it melt. Add both the chocolates and keep stirring until chocolates melt.

3. Add cocoa powder and whisk well. Turn off the heat.

4. Add egg in a bowl and whisk. Add sugar, vanilla, and salt and whisk well. Pour the melted chocolate mixture and keep whisking until well incorporated.

5. Add flour and fold gently. Spoon the batter into the baking dish. Place the baking dish in the oven and bake until firm on top. It should take about 20 minutes. When ready, a toothpick, when pierced in the middle of the brownie, comes out without any particles stuck on it.

6. Let the baking dish cool on your countertop.

7. Cut into 9 equal pieces.

8. To make drizzle: Melt chocolate and shortening in a double boiler. The method is given in the previous recipe. Keep stirring until chocolate melts.

9. Drizzle the chocolate over the brownies.

Canna Chocolate Dipped Strawberries

Preparation time: 5 minutes

Cooking time: 2 minutes

Makes: 6 servings

Ingredients:

- 1 tablespoon canna-coconut oil

- 6 strawberries with stem

- ¾ cup chocolate chips

Directions:

1. Add chocolate chips and coconut oil into a microwave-safe bowl. Cook on high for about a minute. Stir every 12 – 15 seconds until chocolate melts completely.

2. Line a baking sheet with parchment paper. Hold the strawberries with its stem and dip it in the chocolate. Lift it and place it on the baking sheet. Once chocolate sets, it is ready to serve.

Pineapple Upside-Down Cake

Preparation time: 15 minutes

Cooking time: 50 – 60 minutes

Makes: 18 – 20 servings

Ingredients:

- 12 – 14 canned pineapple slices, drained

- 4 cups granulated sugar

- 4 cups cake flour

- 2 teaspoons salt

- 3 tablespoons dark rum

- 4 eggs

- 1 ½ cups canna-butter, at room temperature

- 1 ¼ cups firmly packed light brown sugar

- 2 ¼ teaspoons baking powder

- 1 ½ cups milk

- 3 teaspoons vanilla extract

- 12 – 14 maraschino cherries

Directions:

1. To prepare oven and baking dish: Place rack in the lower third position in the oven and make sure that the oven is preheated to 350° F.

2. Grease 2 pie pans (9 inches) with some oil or butter.

3. Divide equally the pineapple slices and lay them on the bottom of the pan, next to each other, without overlapping.

4. Add ¾ cup canna-butter, 1-cup light brown sugar, and 1 cup granulated sugar into a saucepan.

5. Place the saucepan over medium flame. Keep stirring until butter melts. Remove the pan off the heat and keep stirring until sugar is dissolved completely.

6. Divide the sugar solution equally and pour it all over the pineapple slices.

7. Sift cake flour, salt, and baking powder in a bowl.

8. Combine rum, milk, ¼ cup brown sugar, and vanilla in a bowl.

9. Add remaining canna-butter and remaining sugar into a mixing bowl. Beat with an electric hand mixer, set on high speed until light and creamy.

10. Beat in the eggs, one egg each time. Beat well each time. Add vanilla extract.

11. Set the mixer on low speed and add the mixture of flour and mixture of milk, a little at a time and beat until just combined, making sure not to over-beat.

12. Divide equally the batter among the pie pans, over the, over the pineapple slices.

13. Bake the cakes in a preheated oven at for about 50 – 60 minutes or until a toothpick, when pierced in the middle of the cake, comes out without any particles stuck on it.

14. Switch off the oven and let the pie pans remain in the oven for 10 minutes.

15. Remove the pie pans from oven and let them cool for 15 minutes. Run a knife around the edges of the cake and invert on to 2 plates.

16. Cut each into 9 – 10 slices. Place a cherry on each piece and serve.

17. Store leftovers in an airtight container in the refrigerator. Remove from the refrigerator an hour before serving.

Macadamia & White Chocolate Cookies

Preparation time: 15 minutes

Cooking time: 10 minutes

Makes: 12 – 15 servings

Ingredients:

- ½ cup canna-butter, at room temperature

- ½ cup chopped macadamia nuts

- 6 tablespoons packed, light brown sugar

- ½ cup chopped white chocolate

- 1 ¼ cups flour

- ¼ cup cane sugar

- ½ teaspoon baking soda

- ¼ teaspoon vanilla extract

- ¼ teaspoon almond extract

- ¼ teaspoon salt

- 1 egg

Directions:

1. Add canna-butter, cane sugar, and brown sugar into a mixing bowl and mix with an electric hand blender until creamy.

2. Add egg and beat well. Add vanilla extract and almond extract and beat well.

3. Add flour, baking soda, and salt into a bowl and stir until well combined. You can also sift them together.

4. Add the flour mixture, about 2 tablespoons at a time, and mix well each time.

5. When all of the flour mixture is added, stir in the nuts and chocolate.

6. You need a baking sheet without any grease on it.

7. Scoop out the mixture and drop them on the baking sheet. You should have about 12 – 15 cookies.

8. Place the baking sheet in an oven that has been preheated to 350° F and bake for 10 minutes or until golden brown around the edges.

9. Let the cookies cool on the baking sheet.

10. Transfer into an airtight container and store at room temperature. It should last for 7 – 10 days.

Marijuana Chocolate Chip Cookies

Preparation time: 10 minutes

Cooking time: 10 – 12 minutes

Makes: 25 – 30 servings

Ingredients:

- 4 ¾ cups flour

- 2 teaspoons salt

- 2 ounces butter

- 2 big cups brown sugar

- 4 large eggs

- 2 teaspoons baking soda

- 12 ounces canna-butter

- 1 ½ cups sugar

- 2 teaspoons vanilla extract

- 3 ½ cups chocolate chips

Directions:

1. Sift flour, salt, and baking soda into a bowl.

2. Add canna-butter, brown sugar, vanilla, and sugar into a mixing bowl. Beat with an electric hand mixer set on high speed. Keep beating until fluffy.

3. Beat in the eggs, one at a time, and beat well each time.

4. Add the mixture of dry ingredients and mix well.

5. Add chocolate chips and mix well.

6. You need 1 – 2 baking sheets lined with parchment paper.

7. Scoop out the mixture and drop them on the baking sheet. You should have about 25 – 30 cookies.

8. Place the baking sheet in an oven that has been preheated to 375° F and bake for 10 minutes or until golden brown around the edges.

9. Let the cookies cool on the baking sheet.

10. Transfer into an airtight container and store at room temperature. It should last for 7 – 10 days.

Weed Chocolate Bars

Preparation time: 5 minutes

Cooking time: 10 – 15 minutes

Makes: 4 servings

Ingredients:

- ½ cup canna-butter

- 6 tablespoons powdered sugar

- 6 tablespoons unsweetened cocoa

Directions:

1. Take 2 pots of nearly the same (but not same) sizes such that the smaller one fits inside, the larger pot., The smaller pot should not touch the bottom of the bigger pot. It should fit well inside it.

2. Pour enough water into the larger pot such that it is 1/3 full. The water should not touch the smaller pot. Place the bigger bowl over medium flame. Let the water come to a boil.

3. Add butter into the smaller pot. Place the smaller pot inside the bigger pot.

4. Lower heat to low heat and let the water simmer. Once butter melts, remove the smaller pot from the double boiler.

5. Sift together cocoa and powdered sugar and add into the melted butter. Stir until well incorporated. Take a chocolate bar mold and pour the chocolate mixture into the mold.

6. Once completely cooled, place in the refrigerator for a couple of hours or until set.

7. Unmold and serve.

Vegan Pumpkin Spice Ice Cream

Preparation time: 5 minutes

Cooking time: 0 minutes

Makes: 10 – 12 servings

Ingredients:

- 1 – 2 teaspoons pumpkin pie spice

- 4 cans full fat coconut milk

- 1 teaspoon vanilla extract

- A large pinch sea salt

- 2 – 4 tablespoons maple syrup

- 2 – 4 tablespoons canna-coconut oil, melted

Directions:

1. Shake the cans of coconut milk well and pour into ice cube trays. Freeze until use. If you do not have sufficient ice trays, pour into a pan lined with parchment paper and freeze until firm. You can then chop them into chunks and use as required.

2. Add coconut milk ice cubes, pumpkin pie spice, vanilla, salt, maple syrup, and canna-coconut oil into the food processor bowl. Process until well combined.

3. For soft-serve consistency, you can serve right away.

4. For firm ice cream, transfer into a freezable bowl and freeze until firm.

Weed Vanilla Ice Cream

Preparation time: 30 minutes

Cooking time: 5 minutes

Makes: 8 servings

Ingredients:

- 1 ½ cups white sugar

- 4 ½ cups canna-milk

- 2 cups heavy whipping cream

- 4 teaspoons vanilla extract

Directions:

1. Combine sugar, canna-milk, cream, and vanilla in a saucepan.

2. Place the saucepan over low flame and stir frequently until sugar dissolves. Turn off the heat and pour into a bowl.

3. Add vanilla extract and stir. Cool completely. Cover the bowl with cling wrap and chill for 7 – 8 hours.

4. Add the mixture into an ice cream maker. Follow the instructions of the manufacturer and make the ice cream.

5. You can serve right out of the ice cream maker for a soft-serve consistency. Else transfer into a freezer-safe container and freeze until firm.

6. If you do not have an ice cream maker, pour into a freezer-safe container and freeze for 2 hours. Whisk well and freeze once again until firm.

Banana Marijuana Ice Cream

Preparation time: 1minutes

Cooking time: 5 – 7 minutes

Makes: 20 – 25 servings

Ingredients:

- ½ stick butter

- 10 tablespoons sugar

- 1/8 teaspoon salt

- 6 tablespoons rum

- 0.7 ounce finely ground marijuana

- 36 ounces cream

- 30 ounces bananas, peeled, mashed

- 10 tablespoons honey

Directions:

1. Pour cream into a saucepan and place it over medium flame. When the cream is heated and simmering (it should be hot but not boiling), stir in the marijuana. Mix well and turn off the heat.

2. Add butter, sugar, and salt into another saucepan. Place over low heat to melt butter. Once butter melts, turn off the heat and stir until well combined.

3. Add the cream and bananas into the saucepan of butter mixture and whisk well.

4. Add honey and rum and beat until well combined.

5. Spoon the mixture into a freezer-safe container. Keep the container covered and place in the freezer. After 3 hours, remove the ice cream from the freezer and transfer the ice cream into a chilled bowl.

6. Whisk well. Cover with cling wrap and freeze until firm. 30 minutes before serving, remove the ice cream from the freezer and place in the refrigerator.

7. Serve.

Hash Fudge

Preparation time: 5 minutes

Cooking time: 10 minutes

Makes: 30 – 40 servings

Ingredients:

- 1 ½ cups heavy cream

- 2 teaspoons cornstarch

- 0.21 ounce hash

- 2 tablespoons vanilla extract

- 4 ounces chocolate, unsweetened, chopped

- 6 tablespoons butter

- 4 cups sugar

Directions:

1. Add milk, sugar, chocolate, and cornstarch and into a saucepan and stir. Place the saucepan over medium flame.

2. When the temperature of the mixture reaches 240° F, turn off the heat. Insert a cooking thermometer (also called candy thermometer) to check the temperature. Let the thermometer remain in it.

3. Add butter to a microwave-safe bowl. Microwave on high for a few seconds until the butter melts. Stir in the hash.

4. Continue cooking for another 30 seconds.

5. Pour this mixture into the bowl of chocolate mixture. Do not mix it.

6. When the temperature reaches 110° (make sure to work at this temperature because hurrying will only spoil your fudge), stir

constantly until it gets difficult to move the spatula. It should take around 7-10 minutes. You need to be vigorous with the stirring.

7. Transfer into a baking dish. Let it cool completely. Cut into squares of the desired size. Store in an airtight container in the refrigerator.

8. Hash can be replaced with canna-butter if hash is unavailable.

No-Bake Cannabis Pumpkin Pie

Preparation time: 15 minutes

Cooking time: 0 minutes

Makes: 18 – 36 servings

Ingredients:

For crust:

- 2 cups pitted, chopped dates
- 1 cup shredded coconut
- ½ tablespoon nutmeg or cinnamon
- 4 tablespoons coconut oil
- 5 cups chopped nuts of your choice
- 2 tablespoons pumpkin pie spice
- 1/8 teaspoon salt

For filling:

- 3 cans pumpkin puree
- 3 cups maple syrup
- 4 cups cashews
- 6 tablespoons canna-coconut oil

Directions:

1. To make crust: Add dates, coconut nutmeg, coconut oil, nuts, pumpkin pie spice, and salt into the food processor bowl and process until well combined and sticky, with a few chunks.

2. Take 2 springform pans of about 10 to 12 inches each. Divide the crust mixture among the pans and press it onto the bottom and a little up the sides of the pan.

3. Place the pans in the freezer for 3 hours.

4. To make filling: Make the filling 10 minutes before pouring over the crust.

5. Add pumpkin puree, maple syrup, cashew, and canna-coconut oil into the food processor bowl and process until smooth.

6. Divide equally the filling among the piecrusts. Place the pans in the freezer for about 2 to 3 hours.

7. Remove from the freezer and place on your countertop for a few minutes before serving.

8. Cut into slices and serve. You can get 18 to 36 servings depending on the size of the slices.

Apple Pie

Preparation time: 20 minutes

Cooking time: 40 minutes

Makes: 8 – 10 servings

Ingredients:

For the crust:

- 1 2/3 cups all-purpose flour

- 1 tablespoon sugar

- 2 tablespoons chilled vegetable shortening, cut into cubes

- ½ teaspoon salt

- ½ cup canna-butter, chilled, cut into cubes

- 2 – 3 tablespoons chilled water

For the filling:

- 1 ½ pounds apples, cored, peeled, sliced

- 1 ½ tablespoons all-purpose flour

- A pinch salt

- A pinch ground nutmeg

- ½ tablespoon water

- 1 2/3 cups brown sugar

- 1 ½ teaspoons ground cinnamon

- ½ teaspoon granulated sugar

- 1 small egg, lightly beaten

- ½ tablespoon fresh lemon juice

Directions:

1. To make crust: Combine sugar, flour, and salt in a mixing bowl.

2. Add canna-butter and cut it into it the mixture until small crumbs are formed.

3. Pour chilled water, a tablespoon at a time, and mix well each time. Keep adding the water until you get smooth dough. Divide the dough into 2 portions, one portion slightly smaller than the other.

4. Wrap the dough balls in cling wrap and place in the refrigerator for 30 – 40 minutes.

5. Roll both the portions on the cling wrap itself and place it in the refrigerator until the filling is prepared.

6. Combine apples, brown sugar, and lemon juice in a bowl.

7. Combine flour, salt, cinnamon, and nutmeg in another bowl. Sprinkle this mixture over the apples and stir well. Let the juices release for 10 minutes.

8. Take a large pie pan and invert the bigger rolled dough over it. Peel off the cling wrap. Press the dough on the sides as well as the bottom of the pie pan.

9. Spread the apple mixture over the crust.

10. Remove the cling wrap from the smaller rolled dough and place the dough over the filling.

11. Seal edges of both the top and bottom dough together. Press the edges with a fork if you want to make a design. Make a few small slits on the top dough.

12. Add egg and water in a small bowl and whisk well. This is egg wash. Brush the egg wash over the top crust. Scatter sugar op top and place the pie pan in the refrigerator for 20 minutes.

13. Place the baking sheet in an oven that has been preheated to 425° F and bake for about minutes or until light brown.

14. Lower the temperature to 325° F and bake for 20 minutes or until golden brown on top.

15. Let the pie cool on your countertop for at least 10 – 15 minutes.

16. Cut into wedges and serve.

Cannabis-Infused Chocolate Cake

Preparation time: 10 minutes

Cooking time: 40 minutes

Makes: 12 servings

Ingredients:

- 1 ¼ cups all-purpose flour

- ½ cup canna-butter

- ½ cup boiling water

- 1 egg

- ¼ teaspoon salt

- 1 cup white sugar

- ½ cup buttermilk

- ¼ cup unsweetened cocoa powder + extra to sprinkle

- 1 teaspoon baking soda

Directions:

1. To prepare the oven and baking dish: Spray a baking dish with cooking spray. Sprinkle a little cocoa powder on the bottom of the dish. The oven has to be preheated to 350° F for about 10 minutes.

2. To mix dry ingredients: Add flour, cocoa, salt, and baking soda into a bowl and stir.

3. Add canna-butter and sugar into a mixing bowl. Beat with an electric hand mixer until creamy with a little fluff.

4. Add egg and beat well.

5. Add buttermilk and beat until just combined. Next, pour boiling water into the batter and beat for a minute.

6. Pour the batter into the prepared baking dish. Place the baking dish in the oven and bake for about 25 – 30 minutes. If a toothpick, when inserted in the middle of the cake, has no particles stuck on it when pulled out, your cake is ready. Otherwise, bake it for a few more minutes.

CHAPTER SEVEN

SALAD & SOUP RECIPES

Salads

Cannabis Chicken Salad

Preparation time: 15 minutes

Cooking time: 0 minutes

Makes: 6 – 7 servings

Ingredients:

- 6 large chicken breasts, cooked, cut into bite-size pieces
- ½ cup diced red bell pepper
- ½ cup diced celery
- ½ cup diced onion
- Pepper to taste
- 2 tablespoons chopped fresh rosemary
- 2/3 cup canna- mayonnaise
- Salt to taste

Directions:

1. Add chicken, bell pepper, celery, onion, pepper, rosemary, canna-mayonnaise, and salt into a bowl and stir until well combined.

2. Cover and chill until use. This salad tastes great when chilled.

Thai Mango Salad

Preparation time: 15 minutes

Cooking time: 0 minutes

Makes: 2 servings

Ingredients:

- ½ tablespoon grated, fresh ginger

- Zest of a lime, grated

- 1 tablespoon canna-olive oil

- ½ tablespoon honey

- ½ jalapeño, deseeded, minced

- 1 small cucumber, peeled, diced

- 1 small red onion, thinly sliced

- 1 ripe mango, peeled, chopped

- ¼ bell pepper or any color, thinly sliced

- A handful fresh cilantro, chopped

- 1 tablespoon lime juice

- ½ tablespoon soy sauce

- 1 clove garlic, peeled, minced

Directions:

1. To make dressing: Add ginger, lime juice, soy sauce, garlic, cilantro, lime zest, honey, and canna-oil into a bowl and whisk well.

2. To make salad: Add jalapeño, cucumber, onion, mango, and bell pepper into another bowl and toss well.

3. Pour dressing over the salad and toss well.

Green Leafy Kale Salad with Brown Canna-Butter Vinaigrette

Preparation time: 15 minutes

Cooking time: 10 minutes

Makes: 2 – 3 servings

Ingredients:

For canna-butter vinaigrette:

- 3 tablespoons unsalted butter

- 3 tablespoons sherry vinegar or red wine vinegar

- 3 tablespoons canna-butter

- Salt to taste

For the salad:

- ¼ cup slivered almonds

- ½ pound Tuscan kale, discard hard ribs and stems, cut into bite-size pieces

Directions:

1. Place a pan over medium flame. Add almonds and toast them until you get a nice aroma. Turn off the heat and let it cool.

2. For canna-butter vinaigrette: Place a small saucepan over high flame. Add unsalted butter. Slowly the butter will melt, and in a few minutes, it will have bits of pieces in it and cook until the bites turn brown.

3. Stir in canna-butter. When it melts, turn off the heat.

4. Stir in the vinegar and salt.

5. Place kale in a bowl. Sprinkle almonds on top. Pour vinaigrette over it. Toss well and serve.

Caesar Salad

Preparation time: 10 minutes

Cooking time: 5 – 6 minutes

Makes: 6 – 8 servings

Ingredients:

- 2 heads romaine lettuce, rinsed, pat dried, chopped

- 8 cups bread cubes

- 6 cloves garlic, quartered lengthwise

- Pepper to taste

- ½ cup olive oil

- Salt to taste

For the dressing:

- 1 ½ cups mayonnaise

- ¾ cup grated parmesan cheese

- 2 teaspoons Worcestershire sauce

- 2 teaspoons Dijon mustard

- 2 tablespoons lemon juice

- 3 tablespoons cannabis-infused olive oil

- Pepper to taste

- 6 garlic cloves, minced

- Salt to taste

Directions:

1. To make dressing: Add mayonnaise, half the parmesan cheese Worcestershire sauce, Dijon mustard, lemon juice, canna-oil, pepper, garlic, and salt into a bowl and whisk well. Chill in the refrigerator until use.

2. To make bread croutons, pan-style: Place a large skillet over medium flame. Add olive oil. When the oil is heated, add garlic and stir-fry until brown. Remove garlic with a slotted spoon and place on a plate.

3. Add bread cubes into the pan. Stir-fry until light brown. Remove bread cubes from the pan and place in a bowl. Sprinkle salt and pepper over it. You can also make croutons by baking, given in the next recipe.

4. Place lettuce leaves, bread cubes, and remaining Parmesan cheese in a large bowl. Drizzle dressing over it. Toss well.

5. Serve.

Weed Salad

Preparation time: 10 minutes

Cooking time: 10 – 12 minutes

Makes: 4 servings

Ingredients:

- 4 thick slices bread, cubed

- Salt to taste

- 2 tablespoons lemon juice

- ¾ cup cannabis salad dressing

- ¼ cup grated parmesan cheese

- 6 teaspoons extra-virgin olive oil

- Pepper to taste

- 5 cups mixed lettuce leaves

- 2 cloves garlic, minced

- 2 cups grilled, chopped chicken

Directions:

1. To make croutons, baked style: Place croutons in a bowl. Drizzle oil over it. Sprinkle salt and pepper over it and toss well. Spread it over a baking sheet.

2. Place the baking sheet in an oven that has been preheated to 400° F and bake for 10 minutes or until golden brown.

3. Let the croutons cool on the baking sheet.

4. Transfer into an airtight container and store at room temperature until use.

5. Add salt, garlic, and lemon juice into a large bowl and whisk well.

6. Whisk in the cannabis dressing (few salad dressings are given in Chapter 3, use whatever suits you).

7. Add pepper to taste. Cover and set aside for a while for the flavors to blend in.

8. To make salad: Add lettuce, Parmesan, and chicken into the bowl of dressing and toss well.

9. Divide into plates. Scatter croutons on top and serve.

Strawberry Burrata Salad

Preparation time: 15 minutes

Cooking time: 0 minutes

Makes: 2 servings

Ingredients:

For vinaigrette:

- 1 ½ tablespoons strawberry puree

- ¾ tablespoon raw honey

- 6 tablespoons avocado oil

- 1 ½ tablespoons white balsamic vinegar

- 1/8 teaspoon pepper

- 10 drops CBD oil

For the salad:

- 3 cups packed arugula

- 2 tablespoons slivered, toasted almonds

- ½ cup sliced strawberries

- 2 balls (4 ounces each) burrata cheese

Directions:

1. To make dressing: Add strawberry puree, honey, vinegar, and pepper into a tall plastic glass that comes along with the immersion blender.

2. Using an immersion, blend the mixture until well combined; with the blender on, pour avocado oil in a thin drizzle. Mix until well combined.

3. Add CBD oil and mix well.

4. To make salad: Add arugula, strawberries, and burrata cheese into a bowl and toss well.

5. Add the dressing and toss well.

6. Scatter almonds on top and serve.

Spinach and Orange Salad with Grilled Salmon and Orange Vinaigrette

Preparation time: 20 minutes

Cooking time: 10 minutes

Makes: 8 servings

Ingredients:

For dressing:

- 2/3 cup orange juice
- 2 teaspoons olive oil
- 2 tablespoons soy sauce
- 3 teaspoons grated fresh ginger
- 1 teaspoon black pepper
- 2 tablespoons canna-olive oil
- 2 teaspoons toasted sesame oil
- 4 teaspoons agave nectar
- 1 teaspoon minced garlic
- Salt to taste

For salmon:

- 8 salmon fillets, 1 inch thick (6 ounces each)
- 4 teaspoons pepper
- 4 tablespoons fresh lemon juice

For salad:

- 16 ounces baby spinach

- 2 medium avocadoes, peeled, pitted cut into small cubes

- 1 red onion, thinly sliced

- 8 oranges, peeled, deseeded, cut into segments

- 2 small cucumbers, peeled, diced

Directions:

1. To make dressing: Add orange juice, olive oil, soy sauce, ginger, pepper, canna-oil, sesame oil, agave nectar, and garlic into a small jar. Shake the jar vigorously until well combined.

2. Set up your grill and preheat it to medium heat. Spray the rack with cooking spray.

3. Season salmon with pepper. Drizzle lemon juice over the fillets and place on the grill, with the skin side on top. Cook for 5 minutes. Flip sides and cook the other side for 5 minutes or until the fish flakes easily when pierced with a fork.

4. Peel off the skin from the fillets.

5. To make salad: Add spinach, avocado, onion, orange, and cucumber into a large bowl and toss well.

6. Pour dressing over it and toss well.

7. To serve: Place salad on individual serving plates. Place a salmon fillet on each plate, over the salad, and serve.

Cranberry Walnut Salad

Preparation time: 15 minutes

Cooking time: 0 minutes

Makes: 8 servings

Ingredients:

For apple cider cannabis vinaigrette:

- ½ cup canna-extra-virgin olive oil

- 1 teaspoon Dijon mustard

- ½ teaspoon garlic powder

- ½ teaspoon dried basil

- ½ teaspoon dried oregano

- Salt to taste

- 8 tablespoons apple cider vinegar

- 2 teaspoons light brown sugar

- ½ teaspoon freshly ground pepper or to taste

For salad:

- ¼ cup crushed walnuts

- 4 Fuji apples, cored, thinly sliced

- 2/3 cup dried cranberries

- 8 cups baby spinach

- 2/3 cup crumbled feta cheese

Directions:

1. To make dressing: Add mustard, garlic powder, herbs, salt, pepper, vinegar, and brown sugar into a blender. Blend until smooth.

2. With the blender machine running on low speed, pour canna-oil in a thin stream through the feeder tube. Blend until the mixture is emulsified.

3. Transfer the dressing into a bowl. Cover and set aside for a while for the flavors to meld.

4. To make salad: Add apples, walnuts, cranberries, spinach, and cheese into a bowl and toss well.

5. Pour dressing on top. Toss well and serve.

Canna-Quinoa Salad

Preparation time: 10 minutes

Cooking time: 20 minutes

Makes: 8 servings

Ingredients:

For dressing:

- 2 tablespoons olive oil
- 2 tablespoons canna-olive oil
- 2 teaspoons ground cumin
- 1 teaspoon ground black pepper
- 4 tablespoons fresh lemon juice
- 2 teaspoons minced garlic
- 1 teaspoon salt

For salad:

- 1 cup dry quinoa
- 4 tablespoons olive oil
- 2 cups corn, fresh or frozen
- 2 cups peas, fresh or frozen
- 2 cups frozen edamame
- 2 cans (15 ounces each) chickpeas
- ½ cup chopped red onion
- 2 cups chopped mint leaves

Directions:

1. To make salad: Cook quinoa following the directions on the package. You should have 4 cups cooked quinoa. You can use leftover cooked quinoa as well.

2. Place a large skillet over medium flame. Add olive oil. When the oil is heated, add corn, peas, edamame, and chickpeas and stir-fry for 5 to 6 minutes.

3. Turn off the heat and let it cool completely. Transfer the vegetables into a large bowl.

4. Add quinoa, onion, and mint and toss well.

5. While the quinoa is cooking, make the salad dressing. For this, add olive oil, lemon juice, garlic, salt, canna-oil, cumin, and pepper into a bowl and whisk well. Cover and set aside until the quinoa cools.

6. Pour dressing over the salad and toss until dressing is well combined with the salad.

7. Serve.

West Coast Garden Salad with Cannabis Olive Oil & Dill Dressing

Preparation time: 15 minutes

Cooking time: 0 minutes

Makes: 3

Ingredients:

For dressing:

- 6 tablespoons olive oil

- 3 tablespoons balsamic reduction

- 1 tablespoon canna-olive oil

- Pepper to taste

- 1 tablespoon chopped fresh dill

- Salt to taste

For salad:

- ½ head red lettuce, chopped

- ½ large beet, peeled, cut into cubes

- ¼ cup chopped fresh dill

- ¼ cup salted pumpkin seeds

- 1 medium carrot, peeled, grated

- ½ cup grape tomatoes

- ¼ cup crumbled feta cheese

Directions:

1. To make dressing: Add olive oil, balsamic reduction, canna-olive oil, pepper, dill, and salt into a bowl and whisk well. Cover and set aside for a while for the flavors to mingle.

2. To make salad: Add lettuce, beet, dill, pumpkin seeds, carrot, tomatoes, and feta cheese into a bowl and toss well.

3. Pour dressing on top. Toss well and serve.

Almond Apricot Chicken Salad

Preparation time: 20 minutes

Cooking time: 10 minutes

Makes: 8 servings

Ingredients:

For dressing:

- 2 cups sour cream
- ½ cup canna-olive oil
- 1 cup mayonnaise
- 4 teaspoons grated lemon zest
- 2 teaspoons salt or to taste
- 1 teaspoon pepper or to taste
- 2 tablespoons lemon juice
- 4 teaspoons Dijon mustard
- 1 ½ teaspoons dried savory

For salad:

- 1 package (16 ounces) spiral pasta
- 6 cups coarsely chopped broccoli florets
- 1 cup chopped green onions
- 12 ounces dried apricots
- 5 cups cooked, diced chicken
- 1 cup chopped celery

- 1 ½ cups sliced almonds

Directions:

1. To make dressing: Add sour cream, canna-oil, mayonnaise, lemon zest, salt, pepper, lemon juice, mustard, and dried savory into a bowl and whisk well. Cover and set aside until the pasta is cooked.

2. To make salad: Cook the pasta following the directions on the package. Add past 4 minutes before draining the pasta. Drain the pasta in a colander and place it under cold running water for a few minutes until it cools. Let it remain in the colander for 10 minutes.

3. Add pasta, broccoli, green onions, chicken, and celery into a bowl and toss well.

4. Combine the salad and dressing and refrigerate until use. Keep it covered in the refrigerator.

5. Meanwhile, place a pan over medium flame. Add almonds and toast the almonds to the desired doneness.

6. Divide salad into plate. Scatter almonds on top and serve.

CBD Salad

Preparation time: 20 minutes

Cooking time: 0 minutes

Makes: 4 – 8 servings

Ingredients:

For salad:

- 6 – 8 cups mixed greens or arugula

- 4 thick pieces bacon or smoked seitan bacon, cooked

- 1 cup crumbled goat's cheese

- 2 cups chopped walnuts

- 1 cup fresh or dried cranberries

- 2 apples cored, cut into pieces

For dressing:

- 4 tablespoons CBD oil

- 1 cup olive oil

- 4 tablespoons apple cider vinegar

- 4 teaspoons maple syrup

- 3 tablespoons poppy seeds

- 1 teaspoon salt

- 1 teaspoon ground mustard

- 2 – 3 tablespoons poppy seeds

- ½ teaspoon freshly ground pepper

- 2 tablespoons finely chopped shallot

- 1 tablespoon hemp seeds or hemp hearts

Directions:

1. To make dressing: Add CBD oil, shallots, and vinegar into a bowl and whisk well.

2. Whisk in the maple syrup, mustard, salt, pepper, poppy seeds, and hemp hearts.

3. Cover and set aside for at least 15 minutes.

4. Meanwhile, make the salad by tossing together apple, bacon, greens, and walnuts.

5. Pour dressing over the salad. Toss well and serve.

Multiple Sclerosis (MS) Arugula Goat-Cheese Salad with Cannabis Citrus Vinaigrette

Preparation time: 15 minutes

Cooking time: 0 minutes

Makes: 2 servings

Ingredients:

For salad:

- ½ cup baby spinach
- A handful cauliflower florets, sliced
- 3 tablespoons shredded carrots
- 1/8 cup crumbled goat cheese
- 1 ½ cups arugula
- ½ cup baby kale leaves
- ¼ cup sliced shiitake mushrooms
- 1 tablespoon chopped pine nuts
- ½ clementine, separated into segments

For citrus vinaigrette:

- 1 tablespoon flaxseed oil
- 2 tablespoons canna-extra-virgin olive oil
- Pepper to taste
- Juice of ½ clementine
- 1 tablespoons apple cider vinegar

- ½ tablespoon clover honey

- 1 tablespoon fresh lemon juice

- Salt to taste

Directions:

1. Dry the greens by patting with paper towels or a kitchen towel.

2. Add all the greens, mushrooms, carrot, cauliflower, clementine, pine nuts, and goat cheese into a large bowl and toss well.

3. To make dressing: Combine clementine juice, lemon juice, apple cider vinegar, flaxseed oil, and seasonings into a blender and blend until well combined.

4. With the blender machine running, pour canna-oil in a thin stream and blend until emulsified. If you want thick dressing, pour some more oil.

5. Pour dressing over the salad. Toss well and serve.

Infused Fruit Salad

Preparation time: 15 minutes

Cooking time: 0 minutes

Makes: 4 servings

Ingredients:

- 1 cup sliced strawberries

- ½ cup fresh cherries pitted

- ½ cup blueberries

- 2 oranges, peeled, separated into segments, deseeded

For dressing:

- 4 teaspoons canna-olive oil

- 2 teaspoons basil-infused sesame oil

Directions:

1. Toss together the berries, cherries, and oranges in a bowl.

2. Pour basil-infused sesame oil canna-olive oil over the fruits and toss well.

3. Chill for a few hours. It can be served as it is or as a topping for pancake, waffles, or just over anything.

THC Tuna Salad

Preparation time: 10 minutes

Cooking time: 20 minutes

Makes: 2 servings

Ingredients:

- 2 – 3 droppers THC tincture

- 1/8 teaspoon garlic powder

- ½ tablespoon dried parsley

- A pinch dried, minced onion flakes

- ½ tablespoon grated parmesan cheese

- 1/8 teaspoon curry powder

- 1 ½ tablespoons sweet pickle relish

- 3 tablespoon mayonnaise

- ½ can (from a 7 ounces can) white tuna, drained, flaked

- ½ teaspoon dried dill weed

Directions:

1. Combine THC tincture, garlic powder, parsley, onion flakes, Parmesan cheese, curry powder, sweet pickle relish, and dill in a bowl.

2. Add tuna and mix well.

3. Serve as a filling for sandwiches or over crackers or over lettuce leaves.

Soups

Marijuana Tortilla Soup with Vegetables

Preparation time: 20 minutes

Cooking time: 20 minutes

Makes: 6 servings

Ingredients:

- 2 medium corn tortillas, cut into thin strips, as thin as matchsticks

- 1 ½ - 2 cups cooked chicken or turkey

- ½ cup water

- ½ cup shredded Monterey Jack cheese

- 2 cups chicken or turkey stock

- ½ can (from a 15 ounces can) corn with its liquid

- 1 cup cannabis salsa

- Any other vegetables of your choice (optional)

To serve:

- A handful fresh cilantro, chopped

- Lime slices

Directions:

1. Place tortilla strips on a baking sheet without overlapping.

2. Bake the tortilla strips in a preheated oven at for about 15 minutes or until light brown in color. Turn off the oven and let the tortilla strips cool on your countertop.

3. Once cooled, set aside one-third of the tortilla strips and add the remaining strips into a blender and blend until crushed. Do not grind it to fine powder.

4. Add salsa, chicken, water, stock, and crushed tortillas into a soup pot. Place the pot over medium-high flame. When it comes to a boil, reduce the heat and cook on low.

5. Stir in chicken and corn and any vegetables if using. Cook until vegetables are tender.

6. Add cheese and stir until cheese melts.

7. Ladle into soup bowls. Garnish with retained tortilla strips and cilantro. Serve with lime slice.

Marijuana French Onion Soup Au Gratin

Preparation time: 20 minutes

Cooking time: 60 – 70 minutes

Makes: 8 servings

Ingredients:

- 4 tablespoons unsalted butter
- 2 large sweet onions like Vidalia
- 4 large leeks, use only white and pale green parts
- 2 tablespoons cannabis olive oil
- 12 cups beef stock
- 2 bay leaves
- Salt to taste
- 2 2/3 cups shredded gruyere or Swiss cheese
- 2 tablespoons olive oil
- 4 large shallots
- 2 tablespoons minced garlic
- 1 cup dry sherry
- 2 teaspoons Worcestershire sauce
- 2 teaspoons balsamic vinegar
- 8 slices French baguette (one-day-old bread)
- Pepper to taste

Directions:

1. Place a soup pot over medium flame. Add butter and oil. When it heats, add onion, leeks (white part), and shallots and stir-fry for a few minutes until the onions start to look light brown in color.

2. Lower the flame and cook until onions are caramelized, stirring occasionally. Scrape the bottom of the pot to remove any particles that are stuck.

3. Stir in the garlic and cook for a couple of minutes.

4. Pour sherry into the pot and cook on medium-high heat. Scrape the bottom of the pot to remove any particles that may be stuck.

5. Cook until nearly dry. Stir in canna-oil. Once well combined, pour stock and Worcestershire sauce along with bay leaves. When the soup begins to boil, lower the heat and cook for about 45 minutes.

6. Discard the bay leaves. Add balsamic vinegar. Turn off the heat.

7. Take 8 ovenproof bowls and place them on a baking sheet.

8. Divide the soup among them. Divide equally half the cheese among the bowls, and scatter on top. Place a slice of baguette in each bowl. Sprinkle remaining cheese on top.

9. Set the oven to broil mode. Place the baking sheet in the oven and broil until cheese melts.

10. Serve right away.

Cannabis Chicken Noodle Soup

Cooking time: 30 minutes

Makes: 8 servings

Ingredients:

- 1 pound chopped, cooked chicken breasts

- 8 cups chicken broth

- 3 ½ cups vegetable broth

- 2 cups sliced carrots

- 1 cup chopped onions

- 2 celery stalks, diced

- ½ cup peas

- Salt to taste

- 12 droppers cannabis tincture

- Pepper to taste

- 2 tablespoons butter

- 1 teaspoon dried basil

- 1 teaspoon dried oregano

- 3 cups uncooked egg noodles

Directions:

1. Place a soup pot over medium flame.

2. Add butter. Once butter melts, add onions and celery and sauté until the onions are translucent.

3. Add chicken, carrots, oregano, basil, egg noodles, peas, cannabis tincture, and chicken stock.

4. Cook until the vegetables and noodles are tender.

5. Add salt and pepper to taste.

6. Ladle into soup bowls and serve hot.

Greek Lemon Chicken Soup

Preparation time: 10 minutes

Cooking time: 20 – 25 minutes

Makes: 8 servings

Ingredients:

- 10 cups chicken or turkey stock

- 4 cups cooked rice, divided

- 4 egg yolks

- Salt to taste

- 2 chicken breasts, skinless, boneless

- 2 tablespoons canna-olive oil

- Pepper to taste

- ½ cup lemon juice

Directions:

1. Pour stock into a soup pot and place the pot over high flame. When the stock begins to boil, drop the chicken in it and cook until chicken is well cooked.

2. Remove chicken with a slotted spoon. Let the stock simmer on low heat.

3. When chicken is cool enough to handle, shred with a pair of forks and keep it in a bowl.

4. Add 1 cup cooked rice and 2 cups of stock into a blender and blitz until smooth.

5. Let the blender be running, pour yolks, canna-olive oil, and lemon juice and blend until smooth.

6. Pour into the simmering soup. Stir often. Also, add in the chicken and 3 cups rice and simmer for about 12 – 13 minutes or until you are satisfied with the thickness.

7. Ladle into soup bowls and serve.

Vegan Split Pea Soup

Preparation time: 1 hour and 15 minutes

Cooking time: 60 – 90 minutes

Makes: 3 – 4 servings

Ingredients:

- 2 ½ cups vegetable broth
- 2 cups dried split peas, rinsed, soaked in water for an hour
- ½ tablespoon canna- extra- virgin olive oil
- 1 ½ tablespoon canna-coconut oil
- 4 cloves garlic, finely minced
- 1 bay leaf
- Pepper to taste
- 2 – 3 cups water
- 1 large carrot, cut into bite-size chunks
- ½ large white onion, finely diced
- 1 ½ -2 tablespoons white miso paste or to taste
- ½ teaspoon dried thyme
- Salt to taste

Directions:

1. Place a soup pot over medium flame. Add canna- extra-virgin olive oil. When the oil is heated, add onion and sauté until onions are translucent.

2. Add garlic and sauté until aromatic. Add carrot and split peas. Add salt and pepper to taste.

3. Stir in miso, thyme, canna- coconut oil, and bay leaf and mix well.

4. Stir in broth and water. Let it come to a boil.

5. Lower the heat and keep the pot cover with a lid. Simmer until split peas are cooked. Stir occasionally. Discard the bay leaf

6. If you have an instant pot, you can cook it in it. It is much quicker.

7. Ladle into soup bowls and serve.

Tomato Soup with Carrots and Celery

Preparation time: 15 minutes

Cooking time: 30 minutes

Makes: 8 servings

Ingredients:

- ½ cup canna-butter

- 2 cans (8 ounces each) tomato sauce

- ¼ cup fresh basil

- Salt to taste

- 6 tablespoons butter

- Pepper to taste

- 2 ½ cups chicken broth

- 2 tablespoons chopped oregano

- 3 cups heavy whipping cream

- 30 baby carrots, thinly sliced

- 6 cloves garlic, chopped

Directions:

1. Place a soup pot over a medium-low flame. Add butter and let it melt.

2. Add all the vegetables and stir-fry until tender.

3. Add herbs, broth, and tomato sauce. Raise the heat to medium and cook for about 20 minutes. Turn off the heat. Cool for a while.

4. Blend the soup in batches and pour it back into the pot. Add cream and place the pot over medium flame. Heat thoroughly.

5. Stir in canna-butter. Once butter melts, turn off the heat.

6. Ladle into soup bowls and serve.

Hearty Vegan Winter Vegetable Soup

Preparation time: 20 minutes

Cooking time: 20 minutes

Makes: 5 servings

Ingredients:

- ½ tablespoon olive oil

- ½ medium onion, finely diced

- ½ large carrot, finely diced

- 6 tablespoons red wine

- ½ can (from a 16.5 ounces can) diced tomatoes with its juice

- 1 ¼ cups water

- ¼ small head green cabbage, thinly sliced

- 1 ½ tablespoons soy sauce

- ½ teaspoon dried oregano

- Hot sauce to taste

- ½ large leek, use only white part, thinly sliced

- ½ stalk celery, finely diced

- ½ tablespoon minced garlic

- ½ tablespoon balsamic vinegar or cider vinegar

- 2 cups vegetable stock

- 1 cup chopped, mixed vegetables

- ½ teaspoon pepper

- 0.03 ounce decarbed hash or kief, finely ground

- Salt to taste

Directions:

1. Place a soup pot over medium-high flame. Add oil and let it heat. Add onion and leeks and cook until light brown. Stir in carrots and celery.

2. Cook for about 5 minutes. Stir in garlic and cook for a few seconds until you get a nice aroma.

3. Add wine and vinegar and scrape the bottom of the pot to remove any particles that may be stuck.

4. Stir in the tomatoes, cabbage, and mixed vegetables. Pour stock and water and cook on low until veggies are soft.

5. Add soy sauce, oregano, pepper, salt, and kief and cook for a couple of minutes.

6. Ladle into soup bowls and serve.

Cream of Cannabis Soup

Preparation time: 10 minutes

Cooking time: 20 minutes

Makes: 8 servings

Ingredients:

- 6 cups vegetable stock
- ½ cup chopped red onion
- 2 tablespoons flour
- 4 cloves garlic, finely chopped
- Pepper to taste
- 2 cups chopped broccoli
- 2 cups sliced celery
- 4 cups heavy cream
- ½ cup chopped cilantro
- Canna-butter, to suit your requirements

Directions:

1. Place a saucepan over medium flame. Add canna-butter and let it melt.
2. Add onion, garlic, and celery and cook for a couple of minutes.
3. Add flour and stir constantly for about a minute.
4. Stirring constantly, pour stock into the pot. Add broccoli and celery and cook until tender.
5. Add heavy cream and heat thoroughly.

6. Add pepper and stir.

7. Ladle into soup bowls. Garnish with cilantro and pepper and serve.

Cannabis Tomato Soup

Preparation time: 15 minutes

Cooking time: 5 minutes

Makes: 8 servings

Ingredients:

- 6.6 pounds plum tomatoes, chopped into chunks

- 6 cloves garlic, finely chopped

- 1 teaspoon pepper

- 6 tablespoons balsamic vinegar

- 4 tablespoons Italian spice blend

- 4 handfuls fresh cannabis leaves

- 2 tablespoons canna-olive oil

- Juice of a lemon

- Salt to taste

- 1 red onion, chopped

- 2 handfuls fresh cilantro, chopped

- 2 handfuls fresh dill, chopped

- ¼ cup crumbled feta cheese

Directions:

1. Add tomatoes, salt, lemon juice, Italian spice blend, garlic, and pepper into a bowl and stir well.

2. Cover and set aside for an hour.

3. Add into a blender along with vinegar, garlic, onion, and canna-oil and blend until nearly smooth. Pour into 8 bowls.

4. Place cannabis leaves in a pan. Cover with water (about ¼ inch in height from the bottom of the pan). Let the leaves cook for about 5 minutes.

5. Divide the cannabis leaves among the bowls. Sprinkle dill, cilantro, and feta cheese on top, in each bowl, and serve.

Green Cannabis Soup

Preparation time: 15 minutes

Cooking time: 1 hour and 15 – 20 minutes

Makes: 8 – 10 servings

Ingredients:

- 4 carrots, cubed
- 6 cloves garlic, minced
- 1 purple cabbage, sliced
- 2 stalks celery, sliced
- 2 jalapeño peppers, sliced
- Salt to taste
- 4 sweet potatoes, peeled, cubed
- 1 cup chopped parsley
- 2 red onions, chopped
- 4 leeks, sliced
- 2 cups diced plum tomatoes
- 4 bell peppers, cut into squares
- 6 tablespoons olive oil
- Pepper to taste
- 4 tablespoons canna-butter
- Water, as required

Directions:

1. Place a soup pot over medium flame. Add olive oil and canna-butter and let the butter melt.

2. Stir in onion and garlic. Add cabbage, celery, sweet potatoes, celery, jalapeños, leeks, tomatoes, and bell peppers after the onion turns translucent.

3. Add 6 – 8 cups water or add more if desired. Let it come to a boil.

4. Reduce the heat and simmer for 60 – 80 minutes. Turn off the heat.

5. Ladle into soup bowls. Sprinkle parsley on top and serve.

Weed Ramen Noodles

Preparation time: 5 minutes

Cooking time: 30 minutes

Makes: 2 servings

Ingredients:

- ½ stick butter

- 2 servings Ramen kimchi noodles

- 4 cups water

- 0.1 – 0.14 ounce weed, ground

Extra flavorings: Optional

- Hot sauce

- Cheese

- Oregano

- Chili etc.

Directions:

1. Place a pot over medium flame. Add water and heat until slightly hot but do not boil. Stir in the weed.

2. Add butter and simmer on medium-low flame until the weed turns slightly brown in color. Do not boil. If it starts boiling, turn off the heat for a while.

3. Add the ramen noodles with the seasoning that comes with it and the optional flavorings.

4. Simmer until the noodles are cooked.

5. Ladle into soup bowls and serve.

Vegan Creamy Cannabis-Infused Potato Soup

Preparation time: 20 minutes

Cooking time: 30 minutes

Makes: 6 – 8 servings

Ingredients:

- 1 ½ tablespoons canna-oil

- 2 cloves garlic, minced

- 1 ½ pounds potatoes, peeled, cubed

- 1 ½ cups vegetable broth

- Salt to taste

- ½ large onion, chopped

- 1 ½ medium carrots, peeled, sliced

- ¼ teaspoon dried thyme

- 1 cup full-fat coconut milk

Directions:

1. Place a soup pot over medium flame. Add canna-oil and let it heat.

2. Once the oil is heated, add onion and garlic and cook until soft. Stir often.

3. Stir in potatoes, thyme, carrots, and broth. Once potatoes are cooked, turn off the heat.

4. Blend with an immersion blender until smooth. You can also blend it in a food processor or blender. There should be no lumps of the vegetables, so you need to blend it well.

5. Add coconut milk and blend once again. Add salt and stir.

6. Ladle into soup bowls and serve.

Cannabis Mushroom Soup

Preparation time: 15 minutes

Cooking time: 20 minutes

Makes: 2 servings

Ingredients:

- A handful shiitake mushrooms

- 1 cup heavy cream

- 2 small cloves garlic, peeled, minced

- 1 teaspoon flour (optional)

- Pepper to taste

- Canna-butter, as required

- 1 cup vegetable broth

- 1 small onion, chopped

- ½ tablespoon olive oil

- Salt to taste

Directions:

1. Place a pan over medium flame. Add canna-butter, as per your requirement, and let it melt.

2. Add mushrooms and cook until brown. Transfer into a bowl, including the cooked liquid.

3. Let the pan dry. Add oil and let it heat over medium-high flame. Add onion and cook until translucent.

4. Stir in garlic and cook for a few seconds until you get a nice aroma.

5. Pour vegetable broth and cream. Let it come to a boil, stirring often.

6. If you are using flour, mix it with a tablespoon of water and pour into the pan. Stir constantly until thick. Add salt and pepper to taste.

7. Ladle into soup bowls and serve.

Cannabis-Infused Bone Broth

Preparation time: 10 minutes

Cooking time: 24 – 48 hours

Makes: 8 – 10 servings

Ingredients:

- 2 ½ pounds beef bones

- 1 onion, cut into thick slices

- 1 stalk celery, cut into 1 inch pieces

- 1 large bay leaf

- 2 tablespoons apple cider vinegar

- 1 carrot, cut into chunks

- 2 cloves garlic, peeled

- Salt to taste

- 0.07 ounce cannabis, chopped

- Pepper to taste

Directions:

1. To prepare the oven: The oven has to be preheated to 240° F for about 20 minutes.

2. Place bones, vegetables, and cannabis on a rimmed baking sheet and toss well. Spread it all over the baking sheet and place the baking sheet in the oven for 40 minutes.

3. Transfer the bones and vegetables along with cannabis into a crockpot. Add seasoning, bay leaf, and apple cider vinegar and stir. Pour enough water to cover the ingredients in the pot by 3 – 4 inches (above the ingredients).

4. Set the crockpot on "Low" and cook for 24 – 48 hours.

5. Discard the bones, cannabis, and vegetables. You can strain the broth into a jar

6. Ladle into soup bowls and serve. Store leftovers in the refrigerator. It can last for 7 – 8 days.

Butternut Squash Soup with CBD Drizzle

Preparation time: 10 minutes

Cooking time: 40 – 45 minutes

Makes: 3 servings

Ingredients:

- 1 ½ tablespoons olive oil

- ½ cup cashews

- 3 cups peeled, cubed butternut squash

- 1 tablespoon minced, fresh ginger

- ¾ teaspoon ground coriander

- ½ teaspoon curry powder

- Salt to taste

- ½ teaspoon canna- coconut oil

- 1 medium shallot, peeled, finely chopped

- 2 small cloves garlic, minced

- 2 cups vegetable stock

- ¾ teaspoon ground cumin

- ½ tablespoon maple syrup

- ½ teaspoon ground turmeric

- ¾ cup coconut milk, divided

- 3 eggs, poached

Directions:

1. Place a soup pot over medium flame. Add olive oil and let it heat. Add shallots and sauté for a couple of minutes.

2. Stir in cashews. Once the cashews turn light brown, stir in the garlic.

3. Stir-fry for a few seconds until you get a nice aroma. Stir in butternut squash, salt, and all the spices. Keep stirring until you get a nice aroma.

4. Stir in the stock and maple syrup. When the soup begins to boil, lower the heat and cook covered until squash is fork-tender.

5. Stir in ½ cup coconut milk. Turn off the heat. Let it cool for a few minutes. Blend the soup until smooth.

6. Pour the blended soup back into the pot. Heat thoroughly just before serving. Also, poach the eggs just before serving.

7. Add canna-coconut oil and ¼ cup coconut milk into a bowl and whisk well.

8. Ladle the soup into soup bowls. Trickle the coconut milk mixture on top. Swirl lightly. Top with a poached egg in each bowl and serve.

CHAPTER EIGHT

SNACK RECIPES

Weed Deviled Eggs

Preparation time: 5 minutes

Cooking time: 15 minutes

Makes: 6 servings

Ingredients:

- 3 large eggs

- 2 – 3 tablespoons mayonnaise

- Salt to taste

- Pepper to taste

- Paprika to taste

- 2 tablespoons canna-oil

- ½ tablespoon minced green onion

Directions:

1. Place eggs in a saucepan. Cover with water and place the saucepan over high flame.

2. When it comes to a boil, reduce the flame and let it simmer for 10 to 11 minutes. Turn off the heat.

3. Drain off the water and pour cold water into the saucepan. Let the eggs cool completely.

4. Remove the shells and cut each into 2 halves, lengthwise. Carefully scoop out the yolks and place in a bowl. Keep the whites on a serving platter, with the cavity (yolk) side facing upwards.

5. Add canna-oil, mayonnaise, salt, green onion, and pepper and mash well.

6. Transfer this mixture into a piping bag. Pipe the mixture into the cavities in the whites. Sprinkle paprika on top. Chill until use.

Weez-Its

Preparation time: 5 minutes

Cooking time: 25 minutes

Makes: 2 – 3 servings

Ingredients:

- 2 ounces canna-oil

- 2 cups Cheez-Its

Directions:

1. Add Cheez-Its into a bowl. Drizzle canna-oil over it and toss gently.

2. Line a baking dish with aluminum foil. Add the Cheez-Its into the baking dish and spread it evenly.

3. Place the baking sheet in an oven that has been preheated to 250° F and bake for 10 minutes or until golden brown.

4. Cool completely and serve.

Jalapeno Pot Poppers

Preparation time: 30 minutes

Cooking time: 15 minutes

Makes: 24 servings

Ingredients:

- 2 small links chorizo (optional)

- 1 cup grated mozzarella cheese

- 24 medium jalapeño peppers

- 6 large eggs, beaten

- Salt to taste

- Vegetable oil, to fry, as required

- 1 cup grated Monterrey Jack cheese

- 2 teaspoons dried oregano

- 0.07 ounce decarbed kief or hash, finely crumbled

- 2 cups dried breadcrumbs

- Pepper to taste

Directions:

1. Place a skillet over medium flame. Add chorizo and cook until brown, breaking it as you stir and cook.

2. Add mozzarella and Jack cheese and mix well. Remove from heat and let it cool for a while.

3. Slice off a thin part from the top of the jalapeño peppers. Remove the seeds and membranes. You should be able to stuff the peppers with the filling.

4. Make 24 equal portions of the cheese mixture. Also, make 24 equal portions of the kief.

5. Place a portion of the cheese mixture on your palm. Flatten it slightly and place a portion of kief on it. Bring together the edges of the cheese mixture to enclose the kief. Make it elongated in shape and stuff this inside a jalapeño pepper.

6. Press together the top edges of the pepper so that the filling goes right inside.

7. Repeat steps 5 – 6 and fill the remaining peppers.

8. Place breadcrumbs, pepper, and salt in a bowl and stir.

9. Dip the peppers in egg, one at a time. Shaking off extra egg, dredge in breadcrumbs mixture.

10. Repeat the previous step once again with each pepper and place on a baking sheet.

11. Place a large skillet over medium flame. Pour enough oil such that it is about 2 inches in height from the bottom of the pan. When the oil is well heated but not smoking, 325° F, carefully drop a few stuffed and breaded peppers in the oil. Cook until golden brown.

12. Remove the peppers with a slotted spoon and set aside on a plate lined with paper towels.

13. Fry the remaining peppers in batches.

14. Serve with a dip of your choice.

Marijuana-Infused, 3 Layered Popsicle

Preparation time: 5 minutes

Cooking time: 0 minutes

Makes: 8 – 9 servings

Ingredients:

For blueberry layer:

- 1 cup frozen blueberries

- 1/8 cup full-fat coconut milk

- 1 tablespoon maple syrup or stevia to taste

- ½ teaspoon CBD oil

- 1 teaspoon canna-coconut oil

For coconut milk layer:

- Maple syrup to taste

- ½ can full-fat coconut milk

- For strawberry layer:

- 6 ounces strawberries, chopped

Directions:

1. Add blueberries, coconut milk, maple syrup, CBD oil, and canna-coconut oil into the food processor bowl or blender and process until well combined and smooth.

2. Pour 2 tablespoons of the blueberry mixture into 8 – 9 Popsicle molds. Freeze for about an hour or until firm.

3. Rinse the blender.

4. To make coconut milk layer: Add coconut milk and maple syrup into a blender and blend until smooth.

5. Pour into the Popsicle molds and place the molds in the freezer for 15 minutes.

6. Rinse the blender once again.

7. To make strawberry layer: Add strawberries into a blender and blend until smooth.

8. Pour over the coconut layer. Insert the Popsicle sticks in the molds. Freeze until firm.

Watermelon and Lime Cannabis-Infused Popsicles

Preparation time: 10 minutes

Cooking time: 0 minutes

Makes: 3 – 4 servings

Ingredients:

- 1 ½ cups cubed watermelon, seedless

- 2 teaspoons canna-honey

- 1 ½ tablespoons coconut oil

- ½ tablespoon lime juice

Directions:

1. Add watermelon, canna-honey, coconut oil, and lime juice into a blender and blend until pureed.

2. Pour into 3 – 4 Popsicle molds. Insert sticks in them. Freeze until firm.

Baked Apricot Brie

Preparation time: 15 minutes

Cooking time: 30 – 40 minutes

Makes: 24 – 28 servings

Ingredients:

- 2 Brie rounds (8 ounces each)

- 2 cups canna-butter, melted

- ¼ teaspoon salt

- 1 package (16 ounces) filo dough sheets, thawed

- 2/3 cup apricot preserve

- ½ cup roasted, salted, chopped almonds (optional)

Directions:

1. Cut off the outer rind of the Brie rounds if desired.

2. Add canna-butter into a saucepan. Place over low heat until it melts. Turn off the heat.

3. Unfold the filo dough on your countertop.

4. Pull out 2 filo sheets and place on your countertop, slightly overlapping. You should have a rectangle of approximately 24 x 17 inches.

5. Brush melted butter over the filo sheets.

6. Place 2 more filo sheets over this, similarly, overlapping slightly.

7. Repeat step 5-6. You should have a stack with at least 3 to 4 layers. The topmost layer should be brushed with butter.

8. Place one Brie round in the center of the filo rectangle.

9. Smear 5-6 tablespoons of the apricot preserve over the Brie. You can use orange marmalade instead of apricot preserve or any other preserve of your choice.

10. Sprinkle salt and half the almonds over the Brie.

11. Fold the edges of the filo layers over the Brie to cover it completely. Press together the edges to seal.

12. Do this with the other Brie as well (steps 2 to 11).

13. Place the Brie filled filo's on a lined baking sheet, with the seam side facing down.

14. Place the baking sheet in an oven that has been preheated to 375° F, for 30 to 40 minutes or until golden brown. You may find that Brie oozing out of the pressed edges.

15. Remove the baking sheet from the oven and let it cool for 15 minutes.

16. Cut into slices. It is best served over crackers.

Cannabis-Infused Salted Caramel Popcorn

Preparation time: 10 minutes

Cooking time: 15 minutes

Makes: 2 – 3 servings

Ingredients:

- 1 ½ quarts popped popcorn

- 2 tablespoons butter

- 2 tablespoons honey

- ¼ + 1/8 teaspoon baking soda

- 1 tablespoons canna-butter

- ½ cup dark brown sugar

- 1 teaspoon salt

- ½ teaspoon vanilla extract or maple extract

Directions:

1. To prepare oven and baking sheet: Place a sheet of parchment paper over a large baking sheet. Make sure that your oven is preheated to 225° F.

2. Place popcorn in a bowl

3. Add canna-butter and butter into a saucepan. Place the saucepan over medium flame.

4. Add honey, brown sugar, and ¼ teaspoon salt and keep stirring. Slowly the mixture will start turning light brown.

5. Turn off the heat. Add baking soda and vanilla extract and stir constantly. Quickly pour it over the popcorn and toss well

immediately. You need to be very quick else; the caramel will become hard.

6. Transfer the popcorn onto the baking sheet. Season with ¾ teaspoon salt.

7. Place the baking sheet in the oven and bake for about 30 minutes, stirring halfway through baking.

8. Take out the baking sheet from the oven and let it cool for 5 – 8 minutes.

9. Serve. Leftovers can be stored in an airtight container. It can last for a couple of days.

Monster Munchie Balls

Preparation time: 5 minutes

Cooking time: 5 – 6 minutes

Makes: 6 – 8 servings

Ingredients:

- ¾ cup canna-butter

- 2 tablespoons chunky peanut butter

- 1 tablespoon cocoa powder

- 1 ½ cups rolled oats

- 1 ½ tablespoons honey

Directions:

1. Add canna-butter into a saucepan and place the saucepan over low flame.

2. Stir in peanut butter, cocoa, oats, and honey. Stir constantly.

3. Transfer into a baking dish and spread it evenly. Freeze for 10 minutes.

4. Line an airtight container with parchment paper or wax paper.

5. Make small balls of the mixture (about 1 inch) place in the container. Close the lid and refrigerate until use.

Cinnamon Muffins

Preparation time: 15 minutes

Cooking time: 20 minutes

Makes: 6 servings

Ingredients:

- 1 ½ cups flour

- 1 egg

- 1 cup cane sugar

- 1 ½ tablespoons baking powder

- ¼ teaspoon ground nutmeg

- ½ tablespoon ground cinnamon

- 2 ½ tablespoons canna-butter

- ½ cup butter, melted

- ½ cup milk

- 2 ½ tablespoons shortening

- ½ teaspoon salt

Directions:

1. Grease a 6 counts muffin pan with some cooking spray. Place muffin liners in each cup.

2. Add canna-butter, ½ cup sugar, and shortening in a bowl. Beat until creamy.

3. Beat in the egg, salt, nutmeg, and baking powder.

4. Add a little flour at a time alternating with a little milk, and beat until just combined. Do this until all the milk and flour is added. You may have a few small lumps, but that is alright but do not over-beat.

5. Pour into the muffin cups, equal quantity in each cup.

6. Place the muffin pan in an oven that has been preheated to 350° F for 20 minutes or until a toothpick, when inserted in the middle of the muffins, comes out without any particles stuck on it.

7. Remove the muffin pan from the oven and let it cool for 15 minutes. Remove the muffins from the pan and place on a serving platter.

8. In the meantime, add ½ cup sugar and cinnamon into a bowl and stir.

9. Drizzle melted butter over the muffins. Scatter cinnamon sugar over the muffins and serve.

Weed Mozzarella Sticks

Preparation time: 20 minutes

Cooking time: 10 minutes

Makes: 2 – 3 servings

Ingredients:

- 1 cup canna-milk

- ¾ cup breadcrumbs

- 5 mozzarella cheese string sticks

- 1 egg

- 5 egg roll wrappers

- Oil, to fry, as required

Directions:

1. Whisk together eggs and canna-milk in a bowl. Place breadcrumbs in a shallow bowl.

2. Place one egg roll wrapper on your countertop with the one of the corners towards you.

3. Brush water on the 2 opposite edges (far away from you). Lay a cheese string stick on the corner closer to you and roll the wrapper along with the cheese stick up to one-third of the wrapper. Now fold the left corner inwards and the right corner inwards, over the rolled cheese stick.

4. Continue rolling to reach the corner farthest from you. Press the edges to seal.

5. Repeat steps 2 – 4 and roll the remaining egg roll wrappers over the cheese sticks.

6. Dip the rolled mozzarella sticks in egg, one at a time. Shake to remove excess egg. Dredge in breadcrumb and place on a plate.

7. Place a deep fryer pan over medium flame. Pour enough oil such that it is about 2 inches in height from the bottom of the pan. When the oil is well heated but not smoking, 375° F carefully drop a couple of breaded mozzarella sticks in the oil. Cook until golden brown.

8. Remove the mozzarella sticks with a slotted spoon and set aside on a plate lined with paper towels.

9. Fry the remaining mozzarella sticks in batches.

10. Serve with a dip of your choice.

Weed Potato Chips

Preparation time: 5 minutes

Cooking time: 15 minutes

Makes: 2 servings

Ingredients:

- ¼ cup canna-oil

- 1 tablespoon Kernels popcorn seasoning of your choice

- 1 large potato, peeled, cut into thin slices

- 1 tablespoon salt to be used if you are not using Kernels popcorn seasoning

Directions:

1. Prepare a large baking sheet by lining it with parchment paper.

2. Place the potato slices on the baking sheet in a single layer.

3. Brush canna-oil over each of the chips, on both the sides.

4. Place the baking sheet in an oven that has been preheated to 400° F or until crisp and brown.

5. Take out the baking sheet, form the oven, and let it cool for 5 minutes. Sprinkle Kernels popcorn seasoning over the chips.

6. Serve.

Weed Biscuits

Preparation time: 15 minutes

Cooking time: 12 – 15 minutes

Makes: 20 – 25 servings

Ingredients:

- 1 ½ cups boiled, mashed sweet potatoes

- 3 cups all-purpose flour

- 2 tablespoons baking powder

- ¾ cup canna-butter, unsalted, cold, cut into small cubes

- 2/3 - 1 cup milk

- 4 tablespoons sugar

- 2 teaspoons salt

Directions:

1. Grease a baking sheet by spraying it with cooking spray.

2. Add about 2/3-cup milk and sweet potatoes into a bowl and whisk well.

3. Combine flour, baking powder, sugar, and salt in a bowl. Add butter and cut it into the flour mixture using a fork or a pastry cutter until crumbly in texture.

4. Add the sweet potato mixture and fold. If the dough is very dry, then add some more milk, 1 tablespoon at a time, and mix well each time using your hands

5. Form into smooth dough so add more milk if required.

6. Dust your countertop with some flour. Place the dough on it and roll the dough with a rolling pin. Cut into biscuits using a biscuit cutter.

7. Collect the scrap dough and form it into a ball. Re-roll the scrap dough and cut out some more biscuits. Keep doing this step until there is no more dough left to make biscuits.

8. Gently lift the biscuits and place on the prepared baking sheet in a single layer. Leave at least 1-inch gap between the biscuits. Bake in batches if required.

9. Place the baking sheet in an oven that has been preheated to 425° F for 12 to 15 minutes or until slightly hard and golden brown.

10. Remove the baking sheet from the oven and cool the biscuits completely.

11. You can brush with a little canna-honey or canna-butter on top if desired and serve.

Super Lemon Haze Mexican Guacamole

Preparation time: 20 minutes

Cooking time: 0 minutes

Makes: 8 - 10 servings

Ingredients:

- 8 Hass avocadoes, peeled, pitted, mashed

- 4 small heads garlic, peeled, minced

- Juice of 2 limes

- 2 teaspoons paprika

- 1 teaspoon cayenne pepper

- 2 sweet white onions, chopped into ¼ inch cubes

- 2 cups chopped small cherry tomatoes (¼ inch cubes)

- 2 ounces canna-olive oil

- 2 teaspoons chili powder or to taste

- Cracked pepper to taste

- Sea salt to taste

Directions:

1. Add avocado, garlic, lime juice, salt, spices, onion, tomatoes, and canna-olive oil into a bowl. Stir until well combined.

2. Cover and chill for a while for the flavors to set in.

3. Serve with vegetable sticks or crackers. This can be served chilled or at room temperature.

Pepper Crackers

Preparation time: 15 – 20 minutes

Cooking time: 10 minutes

Makes: 20 bite-size crackers

Ingredients:

- ½ cup flour + extra to dust

- 2 teaspoons fresh, chopped rosemary

- 1 ½ tablespoons olive oil

- ½ teaspoon baking powder

- 2 tablespoons ice water

- ½ teaspoon canna-olive oil

- 1/8 teaspoon salt or to taste

- Pepper to taste

Directions:

1. Combine salt, pepper, flour, and baking powder in a bowl.

2. Add rosemary and stir.

3. Combine ice water, canna-oil, and olive oil in another bowl and whisk well.

4. Pour into the bowl of flour and mix until dough is formed.

5. Knead into smooth dough using your hands. If the dough is hard and difficult to handle, cover the dough with a moist cloth for 5 minutes.

6. Knead until you get smooth dough.

7. Cover and set aside for 1 – 8 hours.

8. Dust your cutting board with some flour. Turn the dough onto the cutting board and roll the dough until very thin. Cut into 20 equal squares.

9. Line a baking sheet with parchment paper. Place the crackers on it in a single layer.

10. Place the baking sheet in an oven that has been preheated to 365° F and bake for 8 to 10 minutes or until light brown.

11. Remove the baking sheet from the oven and cool completely. On cooling, they turn crisp.

12. Serve with a dip of your choice.

Chocolate Crackers

Preparation time: 15 – 20 minutes

Cooking time: 10 minutes

Makes: 20 bite-size crackers

Ingredients:

- ½ cup flour + extra to dust
- 2 teaspoons fresh, chopped rosemary
- 1 ½ tablespoons olive oil
- ½ teaspoon baking powder
- ½ tablespoon sugar
- 2 tablespoons ice water
- ½ teaspoon canna-olive oil
- 1 tablespoon cocoa powder

Directions:

1. Combine sugar, cocoa powder, flour, and baking powder in a bowl.

2. Add rosemary and stir.

3. Combine ice water, canna-oil, and olive oil in another bowl and whisk well.

4. Pour into the bowl of flour and mix until dough is formed.

5. Knead into smooth dough using your hands. If the dough is hard and difficult to handle, cover the dough with a moist cloth for 5 minutes.

6. Knead until you get smooth dough.

7. Cover and set aside for 1 – 8 hours.

8. Dust your cutting board with some flour. Turn the dough onto the cutting board and roll the dough until very thin. Cut into 20 equal squares.

9. Line a baking sheet with parchment paper. Place the crackers on it in a single layer.

10. Place the baking sheet in an oven that has been preheated to 365° F and bake for 8 to 10 minutes or until light brown.

11. Remove the baking sheet from the oven and cool completely. On cooling, they turn crisp.

Lemon Crackers

Preparation time: 15 – 20 minutes

Cooking time: 10 minutes

Makes: 20 bite-size crackers

Ingredients:

- ½ cup flour + extra to dust

- ½ teaspoon grated lemon zest

- 1 ½ tablespoons olive oil

- ½ teaspoon baking powder

- 2 tablespoons ice water

- ½ teaspoon canna-olive oil

- 1/8 teaspoon salt or to taste

- 1/8 teaspoon red pepper flakes

Directions:

1. Combine salt, red pepper flakes, flour, and baking powder in a bowl.

2. Add lemon zest and stir.

3. Combine ice water, canna-oil, and olive oil in another bowl and whisk well.

4. Pour into the bowl of flour and mix until dough is formed.

5. Knead into smooth dough using your hands. If the dough is hard and difficult to handle, cover the dough with a moist cloth for 5 minutes.

6. Knead until you get smooth dough.

7. Cover and set aside for 1 – 8 hours.

8. Dust your cutting board with some flour. Turn the dough onto the cutting board and roll the dough until very thin. Cut into 20 equal squares.

9. Line a baking sheet with parchment paper. Place the crackers on it in a single layer.

10. Place the baking sheet in an oven that has been preheated to 365° F and bake for 8 to 10 minutes or until light brown.

11. Remove the baking sheet from the oven and cool completely. On cooling, they turn crisp.

12. Serve with a dip of your choice.

Sweet, Spicy, and Sativa Mixed Nuts

Preparation time: 5 minutes

Cooking time: 7 minutes

Makes: 4 servings of ¼ cup

Ingredients:

- 1 ¼ tablespoons canna-butter or canna-oil

- ½ teaspoon salt

- 1/8 teaspoon chili powder

- ¼ teaspoon dried rosemary, crumbled

- 1 ½ tablespoons dark brown sugar

- 1/8 teaspoon ground cinnamon

- A pinch cayenne pepper

- 1 cup nuts of your choice, toasted

Directions:

1. Prepare a baking sheet by lining it with parchment paper.

2. Add canna-butter into a skillet. Place the skillet over medium flame. When it melts, add salt, chili powder, rosemary, sugar, cinnamon, and cayenne pepper. Stir constantly until sugar melts.

3. Lower the flame and add nuts. Keep stirring until nuts are well coated with the mixture.

4. Transfer the nut mixture onto the baking sheet. Spread it evenly. Cool completely.

5. Transfer into an airtight container and store at room temperature. It can last for a week.

Cannabis Granola Bars

Preparation time: 25 minutes

Cooking time: 30 minutes

Makes: 8 – 10 servings

Ingredients:

- 3 tablespoons canna-butter
- 1 tablespoon maple syrup
- ½ teaspoon ginger powder
- ¼ cup raisins
- 1/8 cup hazelnuts
- ½ cup bittersweet chocolate
- 3 tablespoons honey
- 1 tablespoon brown sugar
- ½ teaspoon cinnamon powder
- ¼ cup chopped almonds
- 1 ½ cups rolled oats

Directions:

1. Take 2 pots of nearly the same (but not same) sizes such that the smaller one fits inside, the larger pot., The smaller pot should not touch the bottom of the bigger pot. It should fit well inside it.

2. Pour enough water into the larger pot such that it is 1/3 full. The water should not touch the smaller pot. Place the bigger bowl over medium flame. Let the water come to a boil.

3. Add chocolate into the smaller pot. Place the smaller pot inside the bigger pot.

4. Lower heat to low heat and let the water simmer. Once chocolate melts, remove the smaller pot from the double boiler.

5. In the meantime, add canna-butter, honey, cinnamon, maple syrup, sugar, and ginger powder into a small saucepan. Place the saucepan over low flame. Stir frequently until sugar dissolves completely.

6. Turn off the heat. Add raisins, oats, almonds, and hazelnuts and stir until well coated.

7. Prepare a baking sheet by lining it with parchment paper.

8. Spread the mixture on the baking sheet and press it evenly.

9. Place the baking sheet in an oven that has been preheated to 350° F and bake for 25 minutes.

10. Remove the baking sheet from the oven and cool completely.

11. Cut into bars of size 1 ½ x 4 inches. Drizzle the melted chocolate over the granola bars.

Hash Yogurt

Preparation time: 5 minutes

Cooking time: 5 minutes

Makes: 4 servings

Ingredients:

- 2 tubs flavored yogurt of your choice

- Butter or coconut oil, as required

- 0.03 ounce hash, crumbled

Directions:

1. Add butter into a pan. Place the pan over low flame. When butter melts, add hash and stir. Cook for a couple of minutes, stirring often. Turn off the heat.

2. Divide into the tubs of yogurt and stir.

3. Serve.

Hush Puppies

Preparation time: 10 minutes

Cooking time: 20 minutes

Makes: servings

Ingredients:

- 2 cups canna-milk

- 3 cups self-rising cornmeal

- 1 cup self-rising flour

- 1 teaspoon baking soda

- 1 teaspoon salt

- 2 eggs

- Vegetable oil for deep frying, as required

Directions:

1. To mix dry ingredients: Add cornmeal, flour, baking soda, and salt into a mixing bowl and stir.

2. To mix wet ingredients: Add canna-milk and eggs into a bowl and whisk well.

3. Combine the wet and dry ingredients in a bowl. Mix well to form a thick batter. Set aside for 10 minutes.

4. Place a small deep pan over medium heat. Add enough vegetable oil to cover at least 2 inches in height from the bottom of the pan.

5. When the oil is hot, about 370° F (it should not smoke), drop the teaspoonful's of batter in the hot oil. Add as many as can fit in the pan.

6. Fry until golden brown all over.

7. Remove with a slotted spoon and place on paper towels.

8. Fry the remaining in batches in a similar manner.

9. Serve with ketchup or any salsa.

Marijuana Hot Wings

Preparation time: 15 – 20 minutes

Cooking time: 15 – 20 minutes

Makes: 4 servings

Ingredients:

- 1 pound fresh chicken wings, discard wing tips, cut into wingettes and drummets

- ¼ cup red hot sauce

- ¼ cup canna-butter, melted

- Vegetable oil for deep frying, as required

- Ranch dressing to serve

Directions:

1. Pat the chicken with paper towels to dry.

2. Place a small deep pan over medium heat. Add enough vegetable oil to cover at least 2 inches in height from the bottom of the pan. Let the oil heat.

3. In the meantime, combine canna-butter and red-hot sauce in a shallow bowl.

4. When the oil is heated to 375° F (it should not smoke), add chicken wings in the pan of hot oil. Add as many as can fit in the pan.

5. Fry until golden brown all over.

6. Remove with a slotted spoon and place on paper towels.

7. Fry the remaining in batches in a similar manner.

8. Add the chicken wings into the bowl of sauce mixture. Toss well.

9. Serve with ranch dressing.

Peanut Butter Balls

Preparation time: 15 minutes

Cooking time: 20 minutes

Makes: 30 – 40 servings

Ingredients:

- ¾ cup peanut butter

- 2 cups confectioner's sugar

- 1 cup semi-sweet chocolate chips

- ½ cup canna-butter, chilled

- 2/3 cup Graham cracker crumbs

- ½ tablespoon shortening

Directions:

1. To melt chocolate: Take 2 pots of nearly the same (but not same) sizes such that the smaller one fits inside, the larger pot., The smaller pot should not touch the bottom of the bigger pot. It should fit well inside it.

2. Pour enough water into the larger pot such that it is 1/3 full. The water should not touch the smaller pot. Place the bigger bowl over medium flame. Let the water come to a boil.

3. Add chocolate and shortening into the smaller pot. Place the smaller pot inside the bigger pot. Stir occasionally.

4. Lower heat to low heat and let the water simmer. Once chocolate melts, remove the smaller pot from the double boiler.

5. To make peanut butter balls: In the meantime, add canna-butter and peanut butter into a mixing bowl. Mix with an electric hand mixer until well combined.

6. Mix in the Graham cracker crumbs. Make small balls of the mixture of about 1 inch in diameter and place on a baking sheet lined with wax paper.

7. Dip the peanut butter balls in the melted chocolate, one at a time. You can use a toothpick to do so. Place the dipped ball back on the baking sheet.

8. Do this with all the balls. Place the baking sheet in the freezer and freeze until the chocolate sets.

9. Transfer into an airtight container and refrigerate until use.

CONCLUSION

As you just witnessed, there are several different and delicious ways to incorporate Cannabis for medical uses in your food. Stronger doses of cannabis are definitely not recommended. To be safe, go in for smaller doses.

The recipes given in the book are meant to be taken in small doses. You can increase or decrease the dose depending on your personal needs, but beware of the side effects that might ensue with increased dosages. We hope these recipes help you manage your medical problems in a scrumptious manner.

www.ingramcontent.com/pod-product-compliance
Lightning Source LLC
Chambersburg PA
CBHW070052030426
42335CB00016B/1864